# MOSQUITO

*If you would see all of Nature gathered up at one point, in all her loveliness, and her skill, and her deadliness, and her sex, where would you find a more exquisite symbol than the Mosquito?*

—HAVELOCK ELLIS, 1920

# MOSQUITO

## A NATURAL HISTORY OF OUR
## MOST PERSISTENT AND DEADLY FOE

ANDREW SPIELMAN, Sc.D.,
AND MICHAEL D'ANTONIO

*faber and faber*

First published in the USA by Hyperion 2001
Published in the United Kingdom in 2001
by Faber and Faber Limited
3 Queen Square London WC1N 3AU

Printed in England by Mackays of Chatham plc, Chatham, Kent

A CIP record for this book
is available from the British Library

ISBN 0-571-20980-7

2 4 6 8 10 9 7 5 3 1

## NOTE TO THE READER

This book is a collaboration between a journalist, Michael D'Antonio, and a scientist, me. The stories and anecdotes told in first-person singular recount experiences from my career and my travels around the world. I hope these stories, and this book in its entirety, will lead you to respect and, perhaps, admire the mosquito as something more than just a pest or a vector of disease. Few creatures on earth are more worthy of scientific inquiry and human wonder.

—*Andrew Spielman*

# CONTENTS

## ACKNOWLEDGMENTS

All authors depend on those who came before. *Mosquito* stands on a tradition of tropical medicine and entomology that is more than a century old, and we have drawn on it here. We are also grateful to have had access to the papers, documents, and volumes of both the Harvard School of Public Health and the university's Countway Library.

Many people provided us with valuable encouragement and advice. However we are most indebted to Deb Rebelo for her research assistance and to Richard Pollack, Ph.D., for both scientific and literary advice. Odessa Deffenbaugh was instrumental in getting our final manuscript in shape. And our editor, Will Schwalbe, gave us invaluable support and assistance.

# THE TROUBLE WITH MOSQUITOES

It is the kind of everyday event that in the wrong place and time becomes unnerving. You are on vacation. Perhaps you are strolling at dusk along Fifth Avenue in New York. Maybe you've been taking pictures in the bush in Kenya, or you are stepping off a ferry in Hong Kong. In a quiet moment you feel the itch behind your knee. You reach down and touch a hot, raised welt—a mosquito bite—and you wonder:

Do mosquitoes in this place carry disease?

Is an outbreak under way?

What are the odds that the particular mosquito that drained *my* blood left something deadly behind?

The mere fact that we ask these questions demonstrates the power of the mosquito. No animal on earth has touched so directly and profoundly the lives of so many human beings. For all of history and all over the globe she has been a nuisance, a pain, and an angel of death. Mosquitoes have felled great leaders, decimated armies, and decided the fates of nations. All this, and she is roughly the size and weight of a grape seed.

At the dawn of the twenty-first century, the mosquito and the pathogens she transmits command attention worldwide. Each year millions die from mosquito-borne malaria. National economies are locked in isolation as a result, in part, of the same disease. Cities in Europe and the United States battle outbreaks of West Nile virus. Dozens of countries fight yellow fever, dengue, filariasis, and a host of deadly encephalitis viruses.

To meet the health threats that are growing worse in many corners of the world, we must know the mosquito and see clearly her place in nature. More important, we should understand many aspects of our relationship to this tiny, ubiquitous insect, and appreciate our long, historical struggle to share this planet.

When I was a college student in the early 1950s, nothing held more drama than the fight to save the world from the diseases that these creatures carried. The entomologists, parasitologists, and physicians in this corner of science conducted fascinating experiments in the lab. Some traveled to exotic, even dangerous places in brave crusades to save lives. They hacked through jungles, paddled up croc-infested streams, braved contact with hostile peoples, all in service to humanity.

I was not the only undergrad at Colorado College to have this interest. An older student who lived across the hallway in my dorm shared my delight with this field. Every day, Phil Longenecker seemed to announce something fascinating and a little grotesque that he had read somewhere or other. A classic example: the way mosquitoes serve botflies in Central and South America. The size of a bumblebee, the fly seizes a mosquito in midair and glues her own eggs to her captive's abdomen. Later, when the mosquito feeds on a person, the damp warmth of human skin causes the fly's eggs to hatch, leaving maggots to burrow into the new host. Soon, the maggot's breathing apparatus can be seen poking through the victim's skin. Within a week, it's as large as a small olive.

In 1951, I accompanied one of my professors on a train trip to Chicago to attend the organizational meeting of the American Society of Tropical Medicine and Hygiene, an annual event that brings together specialists in tropical medicine from all over the world. The conference would mark the moment when a young man who grew up fascinated by the natural world became dedicated to entomology and public health.

I listened for five days as intensely bright and confident men gave talks on parasites and on mosquitoes in the Americas, Africa, and Asia. A keynote speaker described the valiant effort that defeated yellow fever so that the Panama Canal might be built. Stooped and aged, Joseph LePrince, sanitary engineer, seemed to me as though he were a thousand years old. But though his voice creaked, his tale was spellbinding, filled with heroism in the face of great danger. Before the mosquito fighters prevailed, earthmovers were silenced as armies of men languished in hospitals, and coffins lined the platforms at railway stations,

The mosquito men I met at that meeting—at the time they were all men—seemed intelligent and incredibly daring. I wanted to be one of them. Advanced study at the Johns Hopkins School of Hygiene and Public Health and a tour of duty in the United States Navy led to a professorship at Harvard. In my time I have felt both the thrill of discovery and felt the pangs of failure. But after fifty years, the work continues to bring endless challenges and surprises.

The truth of a career in entomology and tropical public health turned out to be as exciting, to me, as its anticipation. Though I could never approach the heights scaled by such pioneers as Patrick Manson, Ronald Ross, and Walter Reed, I shared in their adventure during a time when everything in science seemed possible.

It may be difficult to love the mosquito, but anyone who comes to know her well develops a deep appreciation. A few species, like the iridescent blue-lined *Uranotaenia sapphirina*, are truly beautiful. All manifest exquisite adaptation to their environment. As a larva, the mosquito feeds and navigates in water. As an adult, she walks on water as well as land. She flies through the night air with the aid of the stars. She not only sees and smells but also senses heat from a distance. Lacking our kind of a brain, she nevertheless *thinks* with her skin, changing direction, and fleeing danger in response to myriad changes in her surroundings.

More than most other living things, the mosquito is a self-serving creature. She doesn't aerate the soil, like ants and worms. She is not an important pollinator of plants, like the bee. She does not even serve as an essential food item for some other animal. She has no "purpose" other than to perpetuate her species. That the mosquito plagues human beings is really, to her, incidental. She is simply surviving and reproducing.

The first part of this book is concerned with the life of the mosquito. It explores her world, with all its dangers. And it examines her adaptation to different environments, including those special niches where mosquitoes and people come together.

The second part of this book addresses the mosquito's intimate relationship with human beings. It is remarkable that centuries ago, a few bright minds actually guessed that there was a connection between mosquitoes and diseases such as malaria and yellow fever. But this concept was so fantastic to the mainstream of science at the time that it was dismissed. As recently as 1870, the idea that a mosquito might kill was considered preposterous. The breakthrough discoveries of the mosquito's deadly qualities were shocking to the general public. They led immediately to an era of ferocious scientific competition. What was learned changed our understanding of disease and altered the course of history.

In its final section, this book turns to the modern era, which has understood the dangers posed by mosquitoes and tried to confront them. Much progress has been made. We now have effective therapies for many mosquito-borne diseases. And we have workable methods for confronting mosquitoes in the environment.

Nevertheless, this insect and the pathogens that it carries have proven to be hardy, clever, and relentless. Today, despite all our technology and science, mosquitoes may pose a risk to health virtually anywhere in the world. In fact, our troubles with mosquitoes are getting worse, making more people sick and claiming more lives, millions of lives, every year. Drug-resistant malaria plagues the tropics. And many regions that were once considered fully cleansed of mosquito-borne pathogens have recently begun to suffer these plagues once again.

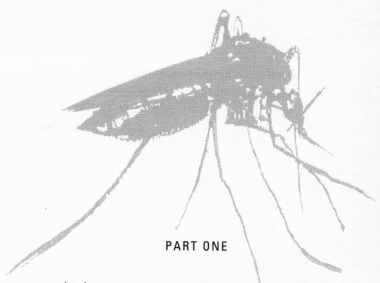

PART ONE

# MAGNIFICENT ENEMY

# 1

# THE MOSQUITO HERSELF

Consider the most common mosquito on earth, one that is likely resting in some dark corner of your very own home or, if you are reading in bed on a warm summer evening, about to issue its faint buzz—do you hear it right now?—in your ear.

This soft, dusty brown insect with a body less than a quarter inch long, is probably *Culex pipiens*, one of more than 2,500 different species of mosquitoes. You've seen her land on your arm. You have caught her just at the end of feeding, her translucent belly swelling red with your blood. At such a moment, you can be forgiven for failing to notice what an elegant and hardy thing she is. But she is.

I have encountered *Culex*—the common house mosquito—in every corner of the world. In one village in Africa, I watched in amazement as thousands coated the walls of a hut, making the surface seem like a living carpet. I have found her beneath the courtyard at Harvard Medical School, in the Paris Metro, and in the gaudy Ginza shopping district in Tokyo. The most impressive infestation I ever saw was at the graveyard of Confucius in China. The stone-lined graves in the cemetery are cracked and filled with water. They are perfect breeding places for *Culex*, which become so numerous at times they look like smoke in the air. Visitors have to cover their mouths to avoid inhaling them with each breath.

Whether she is African, Asian, European, or American, the common house mosquito begins life in essentially the same way, as a tiny egg attached to hundreds of others, which make up a tapered black raft that resembles a half grain of rice floating in still water.

As a young scientist, I recorded how the female *Culex pipiens* creates this little boat, complete with a pointed bow, as she lays and fertilizes her eggs. My observations were made in a lab, where I had used a glass tube with a screen and a rubber hose to inhale a gravid female and deposit her in a test tube filled with water. I then sat, transfixed, for the fifteen minutes she needed to complete her maternal task.

She starts by lighting on the water, crossing her hind legs, and lowering her abdomen. One by one, her eggs are extruded. She twitches so that each one is turned head-down, and placed upright between her legs. Slowly she stacks the eggs, one after another, between her legs. As the eggs move outward into the water, her legs spread to form a little craft with a wide beam and a pointed bow. In the end, approximately 240 eggs float together in the shape of a tiny abandoned canoe.

(Though I was among the first to record the construction of a *Culex* egg raft, I was not the first to witness mosquito egg-laying. Here, the renowned malariologist Paul Russell was a pioneer. Decades before my observations, he and an assistant had crept along the embankment of a rice paddy in India to witness an *Anopheles culicifaces*—a species that carries malaria there—giving birth to eggs that she laid in the water, not in a raft, one by one. Russell and his companion had accomplished their scientific mission with the help of a shiny spoon attached to a stick which, once placed over the water, seemed to attract the mother mosquito who then began ovipositing in flight, like a tiny dive bomber.)

Secure in the raft, each *Culex pipiens* mosquito embryo develops with its head down, resting against a line of fracture encircling its shell. The entire incubation process takes just two days, and at the end, often nearly in unison, the grayish white larvae burst outward by the hundreds. They emerge as long, segmented, wriggling larvae with whiskered heads that hang down in the water for feeding and tails equipped with a breathing tube that takes in air at the surface of the water.

All of the world's insects as well as all arthropods, including crabs and lobsters, are descended from a single segmented ancestor, a wormish creature called an onychophoran. Onychophora, which still crawl the planet today, are made of more than a dozen similar segments, each with a pair of stubby legs. An onychophoran possesses a mouth, which is always open, a simple gut, and an anus at its terminal end.

Though it is sometimes difficult to see, especially in crabs, all arthropods share onychophora's segmentation. It is quite evident in a mosquito larva, which has a wormlike abdomen but is also equipped with a thorax and a well-formed head with eyes and an elaborate mouth. The mouth includes a pair of mandibles for chewing, a pair of laterally mounted maxillae for grasping, and upper and lower lips known as the labrum and the labium. Beyond the head, three segments of the larva's thorax are ornamented with hairs, which the larva uses to stabilize itself in the water. The terminal group of segments bears the air tube, or siphon, through which it breathes. Nearly all larvae drift through life with this tube, located on their rears, poking up against the surface of the water where it can take in air.

If *Culex pipiens* and all the other mosquitoes never developed beyond the larval stage they would still be beautiful and fascinating animals. To the naked eye, they seem to float peacefully, their bodies hanging down from the very surface of the water. Under magnification, they can appear as fierce as any monster in an outer space movie, with large eyes, spiky whiskers, and hungry mouths.

Among the thousands of species of mosquito that have been named there are many variations on this beginning to life. One, *Wyeomyia smithii*, will develop only in the water that collects in North American pitcher plants. In the same bogs where the pitcher plants grow, a mosquito called *Coquillettidia perturbans* has a truly bizarre way of finding its air supply. At birth, the larva of this mosquito swims backward, pressing its siphon against the roots of cattail plants. From the end of the siphon it extends first a kind of tweezer that opens a hole in the cattail root. It then inserts its breathing tube into the hole and uses a series of hooklike appendages to secure itself firmly in place. The siphon brings the larva air from chambers inside the plant's roots. The larva will remain suspended there through this stage of life, and the next.

Mosquitoes are like nesting dolls. At every stage in their development, they already contain the beginnings of organs and muscles they will require in the next. So when the larva begins to evolve into the pupa, it already has inside it all of the pupa's organs. Similarly, the pupa will contain every organ of the flying adult. The insect-within-an-insect phenomenon was observed in the 1920s by one of the grand men of mosquito science, Sir Rickard Christophers. Christophers was a classical scientist who worked well past his one hundredth birthday. Along with his many discoveries, he is remembered for a letter he wrote to the *American Society of Tropical Medicine and Hygiene* taking exception to the publication of his obituary.

As the *Culex* larva reaches about ten days of age, the pupal eye becomes visible as a crescent forming around the round larval eye. Other pupal organs also form, but beyond view, unless we use special techniques. When the larva is about twelve days old, an enzyme begins the process that will lead to the shedding of its outer covering, or cuticle. Slowly, air is taken into the space that separates the old, larva cuticle and the pupa's own cuticle, which is forming beneath it. Eventually, a split develops right down the middle of the thorax. The pupa emerges through the slit, and the old skin of the larva floats away.

With eyes, but no mouth, the pupa appears to be totally different from the larva. It breathes through a pair of tubes—called respiratory trumpets—that are located on the thorax and will become the larva's single siphon. At first its cuticle is soft. But in the course of an hour the cuticle hardens, the abdomen shrinks, and the pupa's body curves into the shape of a comma. Though they don't feed, the waterborne pupae do not just linger in repose. They are mobile creatures with large eyes that are extremely sensitive to a sudden shadow moving over them. If disturbed, they flip their tails and tumble out of harm's way, and then rise again to the surface to breathe.

Our little female house mosquito will spend just two days in the pupal stage, undergoing her transformation from wriggling water creature to flying bloodsucker. If the temperature and other conditions are just right, she will rapidly develop her adult nervous system, muscles, limbs, mouth parts, and other organs.

In the final hours of its development—perhaps in the leaf-clogged rain gutter of your house, or in the forgotten wading pool down the street—the pupa begins the final drama of its brief existence. Right up to the last ten minutes, it retains the ability to flip its tail in pupal fashion and tumble in spurts, like a miniature cannonball. But eventually this movement stops. The pupa stretches its length under the water's surface. The insect inside begins to fill the pupal covering with air until the pressure makes it split open. Slowly the new adult—called the imago—emerges. The back of her thorax comes first, followed by wings, body, legs. Soft and wobbly, she pulls herself up and walks gently on the water. After about a half hour her cuticle stiffens, and she flies away to find a quiet place, protected from wind, rain, and harsh sunlight, where she can rest and complete her development to maturity.

In flight, a *Culex* female's wings will beat between 250 and 500 times per second. The veins that run along the long axis of each wing provide stiffness while allowing for flex, which increases the flow of air. Muscles at the wing's base help it to bow with each stroke, adding to its lifting properties. The whole structure permits a mosquito to fly at speeds of up to three miles per hour and also to hover, though a bit clumsily. All of this flying skill comes in handy whether she is pursuing a blood meal, or approaching a mate.

The exquisite emergence of a newly adult mosquito takes place billions of times each day all over the world. Mosquitoes live eight thousand feet up in the Himalayas and below sea level in the California desert. The eggs of one extremely adaptable mosquito can survive decades of drought in the Sahara until a brief rainfall permits a hatch. (These mosquitoes may hatch already laden by their mother with a virus that can be transmitted to humans, which is one reason that rain in the desert can be a terrible thing.) Above the Arctic Circle, herds of caribou will migrate long distances to avoid the bite of *Ochlerotatus communis*. This mosquito is so aggressive that it feeds on the still warm blood of recently deceased animals.

Although the blood meal—or to be more graphic, the feast on *human* blood—is the one act that we all associate with mosquitoes, it is actually something that the vast majority of mosquitoes will never enjoy. Only a minority of species will deign to feed on human blood, and many of these will turn to other sources as well. And, as is well known, only half of the members of these species, the female half, will practice vampirism at all.

What few people understand is that the one mosquito out of a million that bites a particular ankle is not partaking in an ordinary meal. For the most part, mosquitoes, like butterflies, feed on nectar or other sources of sugar such as rotting fruit. Some species will feed on the honeydew excreted by other insects, and one, in Asia, will actually descend upon a certain kind of ant, insert her proboscis in the remarkably cooperative ant's mouth, and suck out regurgitated nutrients. Nearly all female mosquitoes ingest blood mainly to fuel the production of eggs. But we are getting a bit ahead. Well before she exploits her blood to do her reproductive duty, *Culex pipiens* will have to be impregnated.

Generally neither the male nor the female mosquito is prepared for sex immediately after emerging from the pupal covering. A shocking exception to this rule is an exotic mosquito in New Zealand called *Opifex fuscus*, which develops in pools of salt water along the coast. In this breed, rapacious males stake out the watery places where mosquito pupae are present. As the pupae begin to open, the males will skim across the water like submarine-hunting aircraft. When a pupa rises to break the surface of the water, the male races to the spot. He seizes the pupa, and the force of his attack can cause it to split open. The mosquito inside is helpless, because his or her legs have not yet freed themselves. If the emergent mosquito is a male, the marauder lets go. If it is a female, an event that is essentially a rape ensues, with the male assuring that the female is impregnated in its first moment of adult life.

Among the common house mosquitoes, sex is not quite so violent and dramatic. *Culex* females emerge on their own accord, unmolested, and rest for a few hours to complete maturing after escaping the pupal case. This is an unremarkable process, like tidying up what has already begun. In comparison, the newborn male's situation is downright ridiculous. As he emerges from the cuticle, his sex organs are on the wrong side of his body, and he must wait nearly a day for the terminal segments of his abdomen to rotate around so that everything is aimed in the right direction. At that time he will use his antennae—which are fuzzier, more sensitive instruments than hers—to detect the whining sound of a female's wingbeat. The wing tone of immature males of less than two days' age is quite similar to a female's, which results in some rather unsuccessful encounters among sexually eager, fellow males.

The sounds made by mosquitoes have been studied extensively. Researchers have found that while one kind of mosquito generally ignores the tone of another, the males and the females of the same species are profoundly responsive to the tones of their mates. A well-known story of the powerful attraction the female's sound holds for males involves a power station built in Canada. It malfunctioned repeatedly until engineers discovered that the equipment was being gummed up by thousands of mosquitoes—all males—apparently attracted by the sound the machinery produced.

Once they are about two days old, *Culex* males and females are mature enough to mate. The process begins at dusk or dawn with scores or even hundreds of males forming a dancing swarm in the air near a landmark—called a swarm marker—such as the chimney of a house or a church steeple. (Some species of mosquito will swarm over an animal, or even a person walking in their range.) Though how they choose a marker is not fully understood, the height of an object or the movement, heat, and scent of an animal may be involved.

The *Culex* swarm can be so thick that it might be mistaken for smoke. In many species, generation after generation will swarm in the same location, year after year. This is not because some important bit of information is mysteriously transferred along the bloodlines, but rather because the insects are programmed by nature to perform this way, and males born near the swarm marker will invariably be drawn to it.

Just as the males seek the marker, females will be drawn to the swirling mass of males and fly into it one by one. Often several males will be drawn to a female, but only one will take hold securely enough to remain with her once they've glided to the ground.

Upon landing, the young male crawls beneath the female, and the act can begin. In the most successful encounters, the pair may be so tightly locked together that the male has some difficulty escaping in the end. An unfortunate few males manage to get away only by leaving their sex organs behind. The male also leaves behind in his semen a pheromone called matrone that renders the female much less receptive to future sexual partners. Indeed, it's likely that one sexual encounter is all she'll experience, but males may mate seven or eight times.

As cold—even brutal—as mosquito sex may seem, it is downright romantic compared with the mating behaviors of sea worms, another creature that mates by swarming. One summer night at Woods Hole in Massachusetts, Rachel Carson, the famous author/ecologist, brought me and other students to the end of a small dock at the Marine Biological Laboratory, where she shined a light in the water. A couple dozen male sea worms swam to the illuminated spot and began to circle in their own swarm. When a female entered the swirl, a male responded by entwining himself around her, plunging his tail down her mouth, and depositing his sperm into her gut. And then came the most surprising action of all. The female burst, releasing the eggs and sperm and thereby giving her life for her progeny.

Compared with the female sea worms, female *Culex pipiens* fare quite well. After a sexual encounter, the males fly away with the chance to mate again. For the female, a single minute or so of passion allows her to produce all the fertile eggs she will ever lay. This is because she stores sperm in her own body and dispenses them to fertilize her eggs as they are laid. Thus supplied, she needs just one more ingredient, blood, to nourish the eggs.

Though blood is vital to reproduction for most mosquitoes, nature is almost infinite in her variety, which means that we can make very few absolute declarations about mosquitoes. The fact is, a few kinds of mosquitoes are *autogenous*, meaning they can lay fertile eggs without blood. An even more surprising fact is that some males have been seen sucking human blood. These extremely unusual mosquitoes are gynandromorphs, males that also exhibit some female markings. The gender confusion probably compels them to feed, though they have no use for the blood.

Blood-sucking female mosquitoes of various species search for meals at characteristic times of the day. *Aedes aegypti*, a mosquito that transmits dengue fever from Texas to Argentina, prefers to feed at dusk. The common house mosquito will come out to feed after dark. Other types are most active at midday.

You can get a sense of a mosquito's behavior, and response to stimuli, by imagining how *she* sees *you* during blood feeding. You are standing on a small hill, flapping your arms to ward away the mosquitoes that you sense are buzzing your way. As you swat and flail, the effort increases the carbon dioxide and lactic acid you exhale with each breath. This is good for many mosquitoes, because sensors on their antennae are tuned to these chemicals and others coming from your body. Your scent plume, which is heavier than air, sinks toward the ground. The mosquito flies low, to intersect the plume at its widest part, using her two extended antennae to orient herself in the odor gradient.

But the increase in scent is only one way that you are aiding the enemy. As the mosquito approaches, your waving arms will help stimulate the vision sensors in one or both of her compound eyes. Like a housefly, a mosquito has eyes that are made of hundreds of tubes, each capped with a fixed lens. Called ommatidia, each points outward at a different angle. The lenses of the tube do not focus, so they present an infinite image onto the rhabdom, which is a long photosensitive filament in the center of each ommatidium. Each adjacent, large, and overlapping ommatidium sees a part of the visual field, in full depth.

If we had eyes such as these, there wouldn't be enough neurons in our brains to convert so many images into a picture. Mosquitoes haven't a chance of doing it either. Instead they use their eyes to orient themselves on a fixed point. At night, they can locate distant sources of light such as the moon, or even stars, and fly straight by keeping those lights on a constant bearing. The mosquito who pursues you follows the same strategy, moving to keep all those lenses properly lined up on the point in space you represent. No matter how you move, she adjusts her flight to keep you fixed in her field of vision.

So there you are on this hilltop, waving your arms trying to keep the mosquitoes away, and everything you do is working against you. Your scent tickles the mosquito's antennae. Your movement catches her eyes. Finally, as she gets close, the heat generated by your flailing muscles guides her to the most radiant spot on your body—that bit of flesh not covered by clothing—and your blood.

The mosquito will alight so softly that you may not notice her at all. Before actually feeding, she'll probe your skin as many as twenty times, like a blind phlebotomist who can taste blood with the end of his needle. The probing will be accomplished by a proboscis that is much more than the simple needle most of us imagine it to be. In fact, the proboscis of the female mosquito is much more like the complex devices surgeons snake through a body to perform remote-control surgery.

The business part of the proboscis consists of two tubes surrounded by two pairs of cutting stylets that stick together in a tight bundle. Together they are called the fascicle. The fascicle lies in a gutter-shaped sheath called the labium or lip. When the mosquito lands on the cutaneous oasis from which she'll drink, she applies this apparatus against the skin. With a bit of pressure the labium bends back toward her body as the cutting edges— stylets—of her fascicle break the surface of the skin. (To do this, the stylets slide against each other and split the skin like a pair of electric carving knives.) Once below the surface, the fascicle bends at a sharp angle to begin exploring for blood.

With each insertion, the mosquito attempts to nick a venule or arteriole—vessels much larger than a capillary—and promote the flow of blood. If she doesn't succeed with the first try, she will withdraw the fascicle slightly while leaving it in the original hole, and angle it in a different direction. With each insertion, the fine salivary tube in her fascicle will deliver a chemical that inhibits your body's ability to stop any bleeding that might begin.

Once a mosquito tastes blood, she holds very still, and her sucking pulls the blood venule over her mouthparts. In ninety seconds time she will suck a few micrograms of red fluid—two or three times her weight—up into her stomach. She will stop her feeding when stretch receptors in her belly signal her nervous system that she is full.

After she finishes her meal and she is heavy with the red liquid, a female mosquito must struggle to become airborne. In flight she is a slow target for a swatting hand. But if she escapes, she generally will land on the nearest vertical surface, a wall, a tree, or perhaps a post on a front porch. There she will rest and perform one of the most notable feats of digestion in nature. For forty-five minutes or more, the mosquito will allow her digestive system to draw water out of the precious blood she has stolen and excrete it in the form of urine. If you happen across a mosquito in the middle of this process, you will likely see the pinkish droplets coming out of her anus.

Only after her meal has been processed into lighter solids that are stored for the future will the female mosquito fly away with the stuff to make her offspring. What she leaves behind in the saliva that was pumped into your skin might irritate you or, if your luck is very bad, kill you.

# 2

# AN INSECT'S WORLD

Darting against the blue sky, a hundred feet above the ground, the spotted flycatcher might see nothing more than a flash of reflected sunlight, a shining mirror in the grass below her. As the bird flits lower, the light becomes a pool of polluted water that fills a clogged drainage ditch. She lands on a branch just above the pool's edge. Though clouded by the droppings of cattle and rotting vegetation, the still water gains her attention. She hops down to take a drink.

Countless puddles like this form along the banks of the lower Nile River, all the way south to the Aswan Dam. Completed in 1968, the huge dam, which is visible to astronauts orbiting Earth, ended the annual cycle of flooding and receding waters that had plagued Nile River peoples for all of time. It created cheap and abundant electricity. It brought reliable water supplies to more than a million acres of farmland. And it allowed millions of people to settle deeper into the wadis, shallow valleys where ground now remains perpetually sodden.

The dam improved life immeasurably for local mosquitoes. It created thousands of wet places that produce a harvest of pests as abundant as any other crop in this intensely fertile region. Millions of *Culex pipiens* emerge from these hatcheries every day. They needn't fly far to find convenient sources of blood. It's right there, in nearby farmhouses and apartment buildings. And once the meal is consumed, it's a short jaunt back to the warm, ever present breeding puddle. These tropical *Culex pipiens* focus on human hosts.

When it visits the Nile puddle, the flycatcher might notice a few black specks about the size of caraway seeds floating on this murky puddle. But these are not seeds. These specks are mosquito egg rafts. There is no reason for the bird to remain, so she flits away. As she leaves, one of those seemingly inert, floating specks suddenly comes to life as hundreds of tiny "wrigglers" descend from its lower surface.

What stimulates larval mosquitoes to hatch from their eggs? In many species, especially those that live in more transient waters, the environment offers the hatching cue. Their eggs wait for months or even years in desert, frozen tundra, drying flower vases, or discarded chamber pots. The eggs hatch only when their surroundings become flooded and the water reduces the amount of oxygen available to each egg.

For the eggs of Egyptian *Culex* mosquitoes, the hatching stimulus is an internal series of chemical signals that flow as the embryo matures. These eggs do not rest in a suspended state. They begin to mature as soon as they are laid in water, and they hatch quickly.

Hatching is a dramatic affair. From just below the water's

surface, the egg rafts would seem to be sprouting hundreds of tiny hairs. After a bit of wiggling around, each larva breaks free of the structure. At first they descend an inch or two deeper, almost to the mucky bottom of the puddle. But seconds later, they bob to the surface and orient their tails upward as they reach for air.

Before your anthropomorphic imagination runs amok with images of the *Titanic*'s survivors struggling to breathe in the freezing Atlantic, remember that these larval mosquitoes are born fully adapted to their environment. If all goes normally, each of the just-hatched larvae will simply float, back-end up, until its breathing tube breaks the water's surface. A circlet of valves at the tip keeps the tube safely closed until it meets the air and respiration begins.

Although these larvae will immediately be at home in the water, they do face challenges to their survival. In their first few minutes of life, each must empty the water that fills its tracheolar breathing system and inflate it with air. This network of tubules serves the insect as its lungs. Air diffuses throughout the tracheolar tree, moving passively from the tiny, valved pore at the end of the larva's breathing tube and throughout its body.

The good news for the larvae in this puddle is that no fish are present to gulp them down. This is one advantage of the house mosquito's tendency to breed in filthy, standing water. Other kinds of mosquitoes, which are born from eggs laid at the edges of streams or lakes, must cope with many different kinds of fish. A brood can be rapidly decimated by these voracious creatures. The survivors hide where emergent vegetation creates a complex interface between liquid, solid, and air. Ecologists use the term "intersection line" to describe the suitability of such long-standing waters as breeding sites for mosquitoes.

Snug in their puddles of dirty water, *Culex pipiens* also escape the fate of being eaten alive by other mosquitoes. The wrigglers of *Aedes aegypti* and others born from eggs laid in tree holes, and in man-made containers such as tires, are hunted and consumed by the cannibalistic larvae of *Toxorhynchites*, the largest mosquitoes of all. *Toxorhynchites* feed only on other mosquito larvae that happen to share its breeding site. They are so vicious that once they wipe out all their cousins, some turn on their own brothers and sisters. In the end, they kill just for the sake of killing, letting the uneaten bodies drift away. Often the attacks go on until just a single larva remains.

The *Culex* larvae in our Nile Delta puddle must begin to eat, in order to fuel the growth that will lead them to more than triple their size in about a week's time. This requires a huge amount of food and, as a result, almost constant effort. With their tail-end breathing tubes at the surface of the water, the head-down larval mosquitoes rhythmically wave the rakelike arrays of prominent hairs that surround their mouths. Just as a butterfly's wings create a whispery current of air, the larva's whiskers make a current in the water. This self-made current delivers a never-ending meal to the larva's mouth. The banquet is mainly comprised of microorganisms, the very same animals and plants that you would see if you placed a drop of this puddle water on a slide and looked at it through a microscope. Most of these infinitesimal creatures will nourish the growth of our house mosquito. A few, however, will have a very different effect.

Mosquitoes may encounter in any body of water not only

their food but also various microbes that once scooped into their mouths will lead to infection and death for the mosquitoes. But the fate brought by these infectious agents is not nearly so dramatic or gruesome as the bitter end brought about by certain nematodes.

Ingested with all the other flotsam and jetsam in a puddle, nematodes are innocuous-looking little worms that, once in the larva's gut, begin to get larger. Quickly they grow to two, three, or four times the larva's length. Imagine a worm that fills your body, from your foot to the top of your head, loops around and down and then comes back up again. The nematode grows progressively until, in short order, it consumes its host from the inside out and bursts forth, leaving an empty shell behind.

Sometimes a mosquito's enemy can become our friend. In the Negev Desert of Israel, an entomologist named Joel Margalit once found dying larvae in one of the puddles that occasionally form there after the scarce rains that strike the region. Curious, he brought them to his lab at Ben Gurion University and soon learned that they were infected by a bacterium that could be grown in an artificial culture medium. When a few of the resulting bacterial spores were fed to healthy larvae, they released a complex of toxins that rapidly destroyed the walls of the mosquitoes' guts. Although blackflies are similarly affected, unrelated organisms, amazingly, are completely immune to this bacterial strain.

This host-specific pathogen now provides us with our most useful antimosquito larvicide and is currently applied throughout much of the world. The microbe is known as *Bacillus thuringiensis israelensis*, or more familiarly "Bti," and serves as a universally hailed example of an environmentally friendly insecticidal product.

In our Nile Delta puddle, various microbes, nematodes, and other hazards will claim many of the *Culex* larvae that develop there. Other larvae will fail to thrive if, in a crowded environment, they simply cannot gather enough food. The bodies of the dead fall gently to the bottom of the puddle, adding their bulk to the biomass that fuels the cycle of life. Excessive heat or cold will cause further casualties. By the end of the first day of life, as many as half of the several hundred larvae that emerge from each egg raft will have died.

The surviving larvae are not entirely defenseless. On day two, when a water strider glides over the surface of the puddle, the larger, more developed larvae sense the sudden decrease in the light coming from overhead and dart to the bottom. Down in the depths, their coloration helps them blend with the muck, and they disappear from this insect's view. The water strider will feed only on larvae that are caught on the surface.

When life returns to normal in the puddle, the larvae resume feeding. They will also bend their heads back on their bodies to groom themselves, paying special attention to the opening on their breathing tubes. This behavior, barely evident to the human eye, follows when the larvae curve their bodies into loops. In quiet moments, the warm water is the stage for a graceful sort of larval ballet. As some gently wave a current of food toward their mouths, others bend and twist in their grooming, while hanging by their tails from the water's surface.

The feeding larvae are completely unaware of the clattering

threat that approaches from just beyond the edge of the puddle. But in a moment it arrives, with a metallic beating of wings and a frenzy of violent movement. A small flight of whirligig beetles—shiny, black, armored beasts that are fifty times the size of a larva—crashes into the water like jumbo jets landing on a fishing fleet. (If this scene were staged in a movie, the background music would, no doubt, be the "Ride of the Valkyries.")

The beetles splash down and churn the water. The puddle becomes mayhem. The quickest and most sensitive of the larvae flip themselves toward the bottom of the puddle while their slower hatch-mates are seized by the grasping mouths of their attackers. But even those who escape the initial splashdown aren't safe. Whirligigs are equipped with eyes that are divided in two, with the lower halves positioned to scan beneath the water's surface. The hungry beetles are good swimmers, and they give chase to the escaping larvae. The beetles feed at will, leaving behind only the lucky few.

Over the following several days, more larvae die from disease and predation. On one particularly hot day, the water at the edge of the puddle evaporates, and some of the surviving larvae become stranded and die.

But it is not all death and destruction for the mosquitoes occupying this tiny world. Although the population of common house mosquitoes born from that original raft is reduced, and reduced again, it is also being joined almost daily by new hatchlings that emerge from the egg rafts laid by other adults. After a week, many generations come to share the puddle, and the survivors in our first group are massively larger than they were at hatching. They eventually become ready to enter the pupal stage.

For two or three days, pupal mosquitoes will float quietly just below the water's surface. Though they breathe and retain the ability to detect a moving shadow and tumble themselves out of harm's way, they otherwise tend to lie still. Finally, as they complete their maturation, the adult insects will slowly emerge to stand on the surface film of the water. Their first flight will carry them to dry land for the few hours of rest needed before a tanning process toughens their body walls and wings. But new dangers await.

In the late afternoon, a handful of the recently emerged mosquitoes will manage to fly the few feet from the trunk of a small date tree, which grew there, taking advantage of the abundant water supply. They now cling to the eastern side of the trunk. Because the sun is low in the west, the mosquitoes remain comfortable in the shade. They rest on soft legs, their eyes facing up the trunk, their antennae extended, so that their sensory hairs get a better sense of their surroundings.

Clustered on the trunk about two feet from the ground, none of the day-old mosquitoes notice that an ant is making its way up the tree, its own antennae waving. But the ant senses the one insect whose bad luck finds her closest to the ground. Oblivious to the ant's approach, the mosquito senses danger only after the ant attacks. The ant seeks the nutrient-filled body of the mosquito but manages only to grab her leg. Nature has designed these limbs to break off in a struggle, and when the mosquito strains to escape, she succeeds but leaves her leg behind. Though now an amputee, the ant-ravaged female on the date tree will live and function effectively with her five remaining legs. Any more lost limbs, however, and she might have trouble maintaining her balance, especially if both were lost from the same side.

A smaller brother to the injured mosquito, who rested nearby and was entirely oblivious to the life-and-death drama that took place inches from his resting spot, will not be so fortunate. In a moment, the hungry ant will attack again. This time it will succeed, seizing the unsuspecting male squarely by his soft abdomen. The mosquito's blood, which is clear, flows where the ant's mandibles penetrate his body wall. The ant carries its prey to its underground nest, where it will share its meal.

For the rest of the night, the mosquitoes will rest. Just before dawn, they may fly up to the leaves of the tree, searching for the sugary spots where aphids have deposited their waste product, known as honeydew, which accumulates in large flakes in this rainless region. This is the same sweet substance that the desert peoples who live in this region treasure as candy. The Bible describes these masses of insect feces as "manna" from heaven. Female as well as male mosquitoes rely daily on such a sugar fix, and males eat nothing else.

For our squadron, a meal of manna is followed by a short flight to the wall of a farmhouse. There, the *Culex* from the puddle are joined by others who developed within the courtyard of the building itself. These mosquitoes had a safer, more sheltered beginning, because they hatched into the bucket-size hole lying beneath the spout of the iron-handled water pump that has served the family there for decades. Beetles would never find this pump hole, and the water is rich with the microscopic organisms that thrive on the scraps of food that fall from the plates and cookware that are rinsed beneath the spout.

. . .

As the sun descends beneath the date palms to the west, each male is compelled by his inner chemistry to compete for a mate in the ritualistic mating dance. Hundreds of males will hover frantically above some marker such as the fence post that stands near the tree on which they rested after leaving the puddle. The post stands about shoulder high, and the mating swarm forms just above it. On this day, danger threatens the swarm. Dragonflies have discovered the frantic dance, and to them, it is nothing less than an all-you-can-eat buffet. Iridescent wings flashing green and blue, they buzz through the massed mosquitoes, with their huge forelegs extended. They catch and hold their prey and will devour them in flight.

Many of the swarming males fall prey to the dragonflies and die before nightfall. Many others fail to find a mate. But some live long enough to engage in three or four of these early evening orgies.

For our Nile Valley mosquitoes, night is filled with a new set of perils. Mosquitoes that stray too far from the house, and fly too high above the ground, may be eaten by the bats that hunt them mercilessly, locating them through echolocation. Inside the courtyard of the house, geckos—little lizards that glide effortlessly over walls—also snack on any unwary insects. And then there is the little boy.

Although often the members of the family seem not to notice the many mosquitoes that feed on their blood while they sleep, the littlest boy becomes annoyed by their nocturnal buzzing, the sound that is so essential for attracting a mate. He slaps at them, even when they are resting harmlessly on the wall of the room in which he sleeps. Sometimes he'll catch one that has just fed and is heavy with blood. Her death will leave a red smear— human blood diluted with her own bodily fluids—on the wall. But surprisingly few mosquitoes will succumb to his flailing hands. Their eyes detect the shadow of approaching death. They may also sense the change in air pressure as a hand approaches.

On this night, a *Culex* female will take her drink of blood once the family is asleep. The man she bites, the young boy's father, won't notice, because the mosquito's landing is so soft, and the work she does on his skin is performed so delicately. An arteriole is nicked, and a puddle of blood begins to form in the man's skin. The taste of blood causes the mosquito to stop moving and to drink. Her belly swells, becoming balloon-like.

Once sated, our mosquito's nutrient load is so heavy that she barely can fly. She repairs to the nearby wall to dispose of the vast quantity of water that weighs her down while retaining the more solid nutrients for her eggs. She must then find a hiding place where she can survive the few days required for her to digest her blood meal and for her eggs to mature.

Unlike many other insects, those that feed on blood produce their eggs in batches. This strategy allows them to squeeze the most out of the abundant nutrients in blood. Within a mosquito's stomach, a semiporous envelope forms to enclose the food mass. After a few hours, enzymes are secreted that digest the surface of the food mass, progressively reducing its volume as yolk begins to form in the liver-functioning fat body lying beneath her body wall.

The several days that elapse in the life of our Egyptian house mosquito, extending from the acquisition of her blood meal to the deposition of mature eggs, represent a period that is crucial to the survival of her species. To accomplish her task, she must find a humid place where she can rest safely amid a jungle of marauding predators, while emerging occasionally to eat a bit of manna. Although ants pose the most immediate danger, many predatory bugs, beetles, and reptiles stalk her. Because they are filled with hiding places, brush, piles of leaves, rocks, or firewood offers her the best chance for survival.

If she survives, the mother mosquito must then emerge to lay her eggs. Her home puddle is nearby, and she may happen upon it. If she doesn't, a combination of visual and odor stimuli guide her to a suitable oviposition site. Indeed, the sampling device that is used most effectively in the United States for monitoring the mosquitoes that transmit West Nile virus is a "CDC gravid trap" based on a foul mixture of fermenting materials. Every egg-loaded female is driven to find nutrient-loaded water that will nourish her offspring.

Most females will not live long enough to lay more than one clutch of eggs. But a small number will seek out yet another blood meal and perhaps lay even more eggs. Fewer still will survive to perform a third cycle of blood feeding and reproduction.

Given the predators, human hands, and other hazards they face, it is remarkable that a number of mosquitoes actually reach old age. Their wings fray, just like a flag that has been whipped by the wind. They lose legs and scales to the vicissitudes of life. Upon autopsy, the age of females can be seen in the tiny scars that each clutch of eggs leave on the ovary. (These can be counted under a dissecting microscope.) Other changes are evident in the coiling of the tracheoles that provide air to her stomach or her ovaries or in the appearance of her internal skeleton and the composition of the waterproofing wax that covers her surface.

In the end, time and energy run out. The oldest living mosquitoes of each generation fall to the ground and die. They will have reached the grand old age of five or six months.

# 3

# TIGERS AND TIRES

According to one theory, the male Asian tiger mosquito is a satyr. Greek mythology holds that the satyr existed in a constant state of sexual arousal, and once he had mated with a wandering nymph, she became sterile. This tiger, whose formal name is *Aedes albopictus*, seems to act the same way. Once he arrives in a location, he will mate with any somewhat similar female, those of his own kind as well as those of related species. The local females—who, after all, do the heavy lifting in perpetuating the population—cannot use these alien sperm. And, to make matters worse for their species, they will never mate again. If enough satyrs invade, and the local males are complacent, the newcomers quickly displace the locals.

With his satyr's ways, the tiger has been overwhelming the local *Aedes aegypti* mosquitoes in many parts of the world, including the American South, for the past twenty years or so. This wouldn't matter to us if tigers didn't bite so aggressively and if their range wasn't so great. Indeed, this creature is nicknamed the tiger because it seeks both animal and human blood with a thirsty ferocity. She also happens to look as much like a tiger as a mosquito can, with prominent black and white stripes that can be seen with the naked eye.

People in much of the South will testify to this insect's annoying persistence. And in the laboratory at least, she is also quite capable of carrying the viruses that cause dengue fever, eastern equine encephalitis, West Nile fever, and LaCrosse encephalitis.

So far, the threat of dengue has not yet materialized in the United States. The reason, ironically, lies in the tiger's aggressive nature. The female tiger will attack a broad variety of host animals, thereby "wasting" her limited number of bites in a manner that dilutes her potential for transmitting specific kinds of pathogens to people.

But the public health impact of this newcomer is not yet settled. Because she bites animals and people, the female tiger mosquito could transfer some pathogens that reside in animal hosts to people. This is a realistic worry. But, we wouldn't be worried at all about the tiger if the many countries that import used tires were a bit more concerned about automotive safety.

The story of *Aedes albopictus* in America begins sometime back in the late 1970s, when the United States was importing many used tires from Asia—mainly from Japan and Taiwan—where the practice of recapping worn tires with new treads and putting them back on the highways is illegal. (The tread can peel away at high speeds, causing obvious safety problems.) The tires, which accumulated at tire shops all over Asia, were sealed in cargo containers and loaded on ships bound for America, where they would be recapped and sold to consumers or redistributed elsewhere in the world.

Houston, Texas, is the tire recapping capital of the world, and freighters brought loads of used Asian tires up the Houston ship canal to the port. Sometime in the early 1980s one of those containers was hoisted off a ship and onto a flatbed truck. When it was delivered to a recapping company, the worker who opened the wide metal doors did not notice the few adult tigers that emerged. Of course, the few larvae that may have been sloshing around in the water that remained in some of those tires or the scattered eggs adhering to the tire's inside wall also escaped notice.

Someone should have been on the lookout for such an unwelcome import, because the relationship between tires and so-called tree-hole mosquitoes like the tiger is well known. Tree-hole varieties have evolved to breed in the water that collects in sheltered water-filled cavities like rot holes in trees, coconut husks, bamboo stumps, and the water-filled leaf axils of bromeliads. The rise of human cultures, which introduced artificial containers to the environment, greatly benefited tree-hole mosquitoes. Discarded food containers are a prominent feature of our throwaway society, and they accumulate near houses where a mature mosquito might readily find the sources of blood that she needs for egg production. In the end, people supply the mosquito with all she needs—artificial, water-collecting tree holes and a supply of blood.

But it was not until humankind began saving used tires that mosquitoes found the blackest, most ideally textured, most perfect simulated tree hole imaginable. If you have ever tried to empty the water from an old tire, you will understand why tires make perfect mosquito maternity wards. Roughly four billion old tires are piled up on the American landscape today, with millions more being added every year.

Strangely enough, the first adult tiger to be identified in North America was captured not near a tire in Houston, but in a graveyard in Memphis. On June 2, 1983, Paul Reiter went to a Memphis cemetery to empty traps he had set to capture mosquitoes for a study that he was conducting on St. Louis encephalitis.

Reiter, who worked for the federal Centers for Disease Control, had already collected ninety thousand mosquitoes this way. He couldn't inspect each one before it was packed in dry ice. That task fell to Richard Darsie, one of the agency's experts on mosquito identification. Darsie was accustomed to receiving thousands of ordinary mosquitoes from Reiter. But when he opened one particular package, he was surprised to see a strikingly black specimen with white bands on its legs and telltale silvery stripes down its back. He knew immediately what he was looking at, and got on the phone.

"You'll never guess what I found. . .," he told Reiter.

Reiter had collected the first known example of a tiger in America, but he could only speculate on the path it had followed from Asia to Memphis. He wrote a brief paper that noted the event as an isolated "achievement of modern transportation" and went back to his work on encephalitis. But it wouldn't be long before he learned that the lonely tiger he encountered in a Memphis graveyard was really just one of a new wave of immigrants to the United States.

The Memphis tiger's cousins were noticed first in abundance in 1985 by softball players and picnickers on the east side of

Houston, where the ship canal brings freighters up from the Gulf of Mexico. Responding to complaints about aggressive mosquitoes, the Harris County Mosquito Abatement Program collected samples of larvae in about 160 places. Mosquito taxonomists—experts at identifying species—recognized the tiger in more than half the samples. Not only had *Aedes albopictus* arrived, but it had also become well established. And because many of the tires that arrive in Houston are redistributed worldwide, there would be no stopping its spread.

In the years since the Texas invasion was noticed, health departments across the country have watched for the tiger and, when it arrived, announced it with varying degrees of alarm. In short order this mosquito conquered most of the South and Midwest.

Around the world, other countries that import used tires began tiger-watching too. When a few isolated mosquitoes were found in Australia, the government there began to require that all shipments of tires be fumigated when they arrive in port. (The Australians have suffered periodic dengue outbreaks since 1975; in parts of the tropical Australian north, it is periodically a leading cause of child death.) The tiger also invaded southern Brazil, and its annoying presence has been noted in Cuba, Italy, and various other parts of the western world.

Although the tiger has not yet been held responsible for any outbreaks of human disease in America, her diverse diet adds some troubling elements to the human environment. Tigers can provide the epidemiological bridge that exposes people to the many different kinds of viruses that are perpetuated by more fastidious mosquitoes among various diverse kinds of hosts. Their presence means that we are now somewhat more likely to acquire pathogens that normally reside in birds, deer, and the many kinds of rodents that surround our homes. The danger is increased because this mosquito is a day feeder and commonly bites people who are working in their gardens.

And the tiger is not the only alien mosquito that has become established in the automobile tires that surround our homes. A related tree-hole mosquito, *Ochlerotatus japonicus*, recently was discovered throughout much of the northeastern part of the United States, where surveys were instigated by the developing outbreak of West Nile fever.

Although little is known about their host range and hence their public health importance, *Ochlerotatus japonicus* is another aggressive biter. Mike Turell, a scientist at the U.S. army's Fort Detrick lab, found that these mosquitoes were good laboratory hosts for West Nile fever virus. Residents of infested sites have little doubt that they are notable pests, and they appear to be better adapted to the cold northern winters than Asian tigers.

The story of the tree-hole mosquito that came to Houston illustrates the dynamic competition that plays out among mosquitoes. As it took over territory and squeezed out *Aedes aegypti, Aedes albopictus* performed a kind of Darwinian blitzkrieg. In less than a decade it accomplished a land grab that would have required many centuries in prehistoric times. It all happened with the help of our tires and our oceangoing ships. Before human beings began mucking around with the landscape, mosquitoes had to settle their territorial conflicts on their own. New opportunities for such introductions occur with increasing frequency as airplanes and ships become larger, faster, and more economical.

All mosquitoes are descended from insects that probably

emerged during the Jurassic period, and probably fed on dinosaurs. You might recall that Michael Crichton's novel *Jurassic Park* suggests that dinosaur DNA might be extracted from a blood-filled mosquito from that era that had been trapped in amber. It's amusing to note, however, that the amber-encased mosquito depicted in Steven Spielberg's film treatment of the book appears to be a *Toxorynchites*, a giant mosquito that is one of a few mosquitoes that will never drink blood. Its mouthparts are not up to the job. It is also one of the few mosquitoes that reproduces without such nutrient.

Eons ago, in the real Jurassic world, mosquitoes in different regions began the long process of adaptation to their local environments. Survivors were those insects best able to cope with whatever water, food, temperature, and other environmental conditions that may have been present.

In the Arctic, evolution permitted the rise of such mosquitoes as *Aedes communis*, a mosquito that lays its eggs in tundra wetlands. The eggs survive subzero winter quite nicely. When the ice thaws and the water rises, they hatch almost instantaneously. The larvae become pupae and develop into flying adults just in time to feed on migrating caribou. No one who has been in their range during their short feeding period will ever forget the clouds of mosquitoes that interfere with breathing and cover every exposed inch of an intruder's body. They'll attack the nostrils, mouth, and ears. Out on the tundra, it is not unusual to discover the rotting carcass of a caribou that has been exsanguinated—drained of blood—by these marauding creatures.

The Arctic's highly specialized mosquitoes like *Aedes communis* face little outside competition for their environmental niche. They are protected by their harsh environment; even if an alien mosquito were to arrive, it would find the climate so hostile that surviving the winter would be impossible.

Another good example of adaptation is a mosquito that I encountered in the early 1980s while surveying for mosquitoes with a team in Egypt. We were working northward, along the Red Sea coast toward the Suez Canal. Because the region is so arid, our collecting was largely limited to the horrendously putrid little waste pits that collected sewage behind the homes which lined the road we traveled.

Near an uninhabited promontory named Ras Shukheir—a place where the imagination sees Moses standing as he parted the Red Sea—lies a marsh where water had accumulated in a series of oval potholes that had formed in the rock. Each pothole was about two or three feet across, and they were scattered in a pattern that looked so much like giant, human footprints that we joked that Moses must have been a very large man.

The potholes had become so saturated with salt that white cakes of the stuff floated on their surfaces. There, in this forbidding brine, lay many larval mosquitoes. They were anophelines and were well known to my Egyptian colleagues. A recent study had identified them as *Anopheles stephensi*, an important transmitter of malaria in the Indian subcontinent, Iran, and the Arabian peninsula; it was the only known infestation of this dangerous mosquito to be documented in Africa.

A debate soon arose over the precise identity of these mosquitoes. I had seen *Anopheles stephensi* in other places occupy a niche that differed greatly from this barren site. They developed in fresh water, not salty water. Furthermore, *Anopheles stephensi* is closely associated with humans. They feed largely on human blood. What could these lonesome mosquitoes be eating? An oil well was evident in the distance, but no people lived within many miles of Ras Shukheir. To make matters even more puzzling, these larvae didn't move quite the way I remembered *Anopheles stephensi* moving. Perhaps they had been misidentified.

New species are identified by comparison and testing. In this case, larvae were collected and raised in the lab, where it was discovered that they could reproduce without feeding on blood. In all of mosquito history, I knew of just one report of an anopheline that reproduces this way. This added to the evidence that the Egyptian insects were a previously undescribed mosquito, one that was adapted to an amazingly severe environment, a home that would be far too hostile to support the survival of any competitor. It was their niche, and theirs alone.

In the end, there was little doubt that the salt-pool mosquito we found near Ras Shukheir represented an unnamed species. A subsequent publication on its peculiar physiology used the name *Anopheles sp. nr. salbaii*, a convention designed to avoid assigning a formal name. *Anopheles salbaii* had been described from a similarly desertlike site, far to the south in the Ogaden desert of Somalia. A third member of this group of similar but different species infests the Danakil Depression of Ethiopia.

No one followed up on the formality of assigning a name to this population, which is unusual because identifying a species can be a boon to the ego. The "namer's" own name is cited as part of the formal Latin taxon designator, thereby earning for her or him a bit of immortality, assuming that no future taxonomist discovers a basis for changing the designator and possibly substituting her or his own claim. To the uninitiated, this process has a certain Indiana Jones aura. The intrepid scientist hacks through the jungle, and suddenly, there it is! In reality, finding the new creature is just the beginning of a process that can be fraught with intrigue and conflict.

This became clear to me in 1963, on Grand Bahama Island, when I observed what appeared to be a new mosquito. At the time, the U.S. National Museum—also called the Smithsonian—held the greatest mosquito collection in America and its curator was the highest authority on mosquito species. This taxonomist received my samples and documentation and issued a surprising response: This was not a new mosquito.

According to the museum expert, a specimen of the very same mosquito had been discovered and described, decades before, in the Dominican Republic. As a young scientist, I deferred to the curator's judgment. I published a complete redescription of the species, using the previously assigned name. So much for the thrill of discovery.

But the story continues. Later, a scientist at the University of California at Los Angeles raised adults from the same mosquito's eggs. He challenged the museum taxonomist's judgment. He assigned it a new name, which, in the end, was accepted, and he was rewarded with his own tiny bit of immortality. These mosquitoes subsequently became established near Miami, where they have joined in the battle for the tree-hole/small container habitat with *Aedes aegypti* and *Aedes albopictus*.

The confusion and conflict involved in naming a new species may surprise people outside the field. But the truth is, even the word *species* is a fuzzy concept. A species is conceived of as a group of genetically unique organisms that share genes mutually but exclusively with others of its own kind. But the gray zone in this definition is so broad that dispute is endless.

In the world of insect science, we have a loose agreement on what constitutes a species. First, there are differences in appearance. Though most people have trouble telling one mosquito from the next, the trained eye sees great variety in size, color, the patterns of color, and the overall shapes of mosquitoes. Antennae, legs, mouthparts, even the genitals of mosquitoes vary widely in their size and construction. Beyond a mosquito's appearance, taxonomists also consider its behavior, adaptation to environment, and even its chromosomal banding patterns. Comparisons of DNA relatedness provide an ultimate test of these relationships.

The subject becomes even more complex when we consider that species are divided once again into subspecies. The variation within a single species can be seen on a global scale when the common house mosquito is considered. To the north or the south of the twenty-third parallel we find one kind of *Culex pipiens*, and toward the equator, we encounter another. Along the latitudinal border, roughly fifty miles in each direction, are regions in which hybrids appear. These subspecies can best be distinguished by examining the structure of the male's "aedeagus," his penis. I have found that tropical, but not temperate, males direct a pair of daggerlike projections toward their mates. In spite of this seeming hazard, mating proceeds freely. The tropical *Culex pipiens quinquefasciatus* most frequently feeds on people, and the temperate *Culex pipiens pipiens* on birds. Their three-part name serves to separate these spatially isolated subspecies.

Subgroups may also coexist in a local environment. In Mali, a mosquito scientist named Yeya Toure studied mosquitoes in the village of Mopti seeking to find ways to reduce the high levels of malaria in the area. He found as many as six more or less isolated populations of *Anopheles* mosquitoes that transmit the infection. The diversity suggests that interventions may be more difficult to carry out than expected.

No consideration of the mosquito's remarkable ability to adapt and specialize would be complete without mentioning that identical mosquitoes that share the same gene pool may nevertheless form exclusive little tribes which refuse to mix with outsiders. Indeed, a common house mosquito from under the grandstand at Boston's Fenway Park may be unable to fertilize a female who resides in New York's Yankee Stadium, and this has nothing to do with the age-old team rivalry.

The reason some such pairings fail has been attributed to *Wolbachia pipientis*, a kind of bacterium first discovered in the reproductive organs of mosquitoes captured in the 1920s in a storm sewer near Harvard Medical School. These *Wolbachia* bacteria—named after a famous Harvard microbiologist of that time—differ genetically among many mosquito communities. Ralph Barr, of the University of California at Los Angeles, discovered that the difference, slight as it may be, makes it impossible for the stranger who might be passing through to fertilize a local female. Cytoplasmic incompatibility kills the sperm, thereby resulting in an infertile egg.

How likely is it that a stray mosquito will wander far from home? Mosquitoes have been captured in the sky at altitudes far higher than five thousand feet, which means that some wind-blown individuals might travel hundreds of miles if caught in a storm or prevailing breezes. But under more typical conditions, some mosquitoes will travel up to seven or eight miles to feed. This was established by some ingenious experiments in the Netherlands during the land-reclamation program in the Zeider

Zee. Water in the newly drained areas was too salty to support local mosquitoes, so any that were caught would have to be flying in from the nearest source of freshwater. To attract mosquitoes, Dutch scientists built pig sties on the freshly reclaimed land and then measured the distance to the nearest mosquito-friendly area. They found that they could still attract mosquitoes in small numbers from a distance of ten kilometers (more than six miles).

In the modern world, mosquitoes are aided in their competition for territory by our airplanes, ships, trains, and trucks. As transportation makes the world smaller and much more interconnected, we are required to track those species that are vectors of disease and do our utmost to keep them out. Australia sprays tire shipments; other countries fumigate the cabins of aircraft arriving from abroad, even as the passengers occupy their seats.

Inevitably, when a dangerous mosquito turns up where it's not supposed to be, nations come into conflict. In the late 1970s, after years of being free of dengue, Cuba endured a horrible outbreak that disabled many of the workers who were to harvest the sugarcane crop. The debilitating illness and loss of life were devastating. And millions of dollars worth of sugarcane rotted in the fields because workers were not fit to harvest it. The Cuban government blamed the American Central Intelligence Agency for introducing the virus strains that caused the crisis, but they more likely came via trade with the African nations that Cuba was involved with militarily.

While exploiting the dengue outbreak for propaganda purposes, Cuba also responded with a massive war on *Aedes aegypti*. As an island nation with an authoritarian government, Cuba was at that time the perfect place to mount such a campaign. In a matter of months, the residence index—the number of homes where *Aedes aegypti* was found—dropped below one per thousand, far lower than the threshold at which disease is readily transmitted.

Though it was stopped in short order, the Cuban dengue outbreak had a ripple effect in capitals across Central and South America. The accusations against the CIA were preposterous, but they pointed to the fact that while the dengue mosquito was absent in most of the region, it had been allowed to flourish in America and was quite common in port cities such as Miami, New Orleans, and Galveston, which traded with Latin America. Within the Organization of American States, a forum where hemispheric issues are debated, country after country demanded that the United States take action against *Aedes aegypti*. Eventually the OAS approved a formal resolution calling on Washington to attack the mosquito.

Bowing to international politics, U.S. authorities funded a program that would go after the mosquito throughout the South, despite the fact that it was causing no dengue outbreaks. (The United States had the vector—the mosquito that transmits the dengue virus—but not the pathogen.) If America had been Cuba, the initiative might have succeeded. But instead, people in many local communities challenged the intrusions onto private property, and in one locale after another, the program was tied up in the courts. The cost of the legal battles drained the money allocated for the program; in the end, it was abandoned. Thanks to local politics, *Aedes aegypti* survived in North America, even as it was being squeezed out of its habitats to the south.

. . .

For tens of thousands of years before humans evolved, mosquito species held to their territories without much change. (*Culex pipiens* left Africa and spread around the world only three hundred years ago.) The early human beings, few in number as they were, had little impact on the mosquitoes' habits, and the mosquitoes of very few species actually fed on human blood. This is because a rain forest where humans have not invaded is very stable, and the local species have evolved to live many stories up, in the jungle canopy. They lay their eggs in tree holes and bromeliads and feed on the blood of birds and monkeys.

High in the trees, these mosquitoes can function as disease vectors. Occasionally they will acquire microbes and transmit them to the local primates. In South America, a mosquito called *Hemagogous spegazzini* can cause a full-blown outbreak of yellow fever among howler monkeys. When this occurs, the forest turns eerily silent as the monkeys die off. Foresters are known to avoid cutting trees in areas where silence reigns because a felled tree might bring mosquitoes, and yellow fever, down to ground level.

As a rule, when we encounter mosquitoes in wild places, their presence suggests that a human settlement is nearby. I observed this once while on vacation in Puerto Rico. Near the top of the mountain called El Yunque in the island's rain forest preserve, I was suddenly attacked by mosquitoes. I peered through the leaves to find that some houses had been built on the mountainside, nearby.

Human settlements exert enormous pressure on mosquito populations, encouraging their evolution and adaptation. The tree-hole mosquito that learns to love old tires is just one example. In nineteenth-century Boston, for example, a tidal estuary located around the region now known as Back Bay was filled in as part of an extensive public works program that employed thousands of immigrants who had fled famine in Ireland. Hills were leveled and tons of rock and earth were used to fill much of the river from which the British had attacked Bunker Hill in 1775. The laborers built shanties so that they could live near their poorly drained place of work. The local malaria vector, *Anopheles quadrimaculatus*, proliferated in the resulting grassy pools that formed there and enjoyed an intimate but damaging relationship with these and other residents of Boston.

Other kinds of mosquitoes, most notably the common house mosquito, became fully domesticated as they adapted to man-made environments. *Culex pipiens* breeds in the water that collects in window wells, basement sumps, and other places we provide for its comfort. It spends winters in warm basements or attics. And it sometimes feeds on us.

As human settlements expanded, mosquitoes of many species learned to enjoy our blood and to live in and around our homes where they had regular access to their source of nutrition. Because only the most alert mosquitoes escaped our efforts to squash them, natural selection allowed for these domestic and peridomestic (which means living near human habitats) species to become most agile, sensitive, even delicate in their feeding. It is no accident that the common house mosquito is a nervous little thing. If she weren't, she'd never survive.

Though the mosquito's evolution toward slap-escaping tendencies makes her ever more annoying, the accomplishment pales when set next to her other adaptive accomplishments. These would be revealed to humankind only in the twentieth century, at the precise moment when we believed she could be conquered at last.

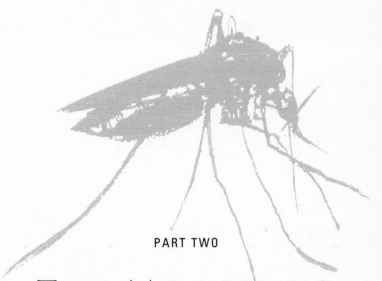

PART TWO

# THE MOSQUITO
# AND DISEASE

# 4

# AGENT OF HISTORY

*When unable to defend herself by the sword*
*Rome could defend herself by means of the fever.*
—The poet Godfrey of Viterbo, 1167

The ancient Romans built a temple on the Palatine Hill honoring the fever goddess. They prayed for mercy from the mysterious, killing illness that appeared every summer. But mercy was often in short supply, and the *campagna* (countryside) was often so malarious that travel there was notoriously dangerous. The sweats, chills, and swelling of the spleen that characterize malaria were widely recognized by Rome's doctors, along with the fact that outsiders were most vulnerable to the disease.

As malaria persisted, and became endemic, the local population developed immunity. But any outsider who ventured there in the hot months risked sickness and even death. Pope Gregory the Great (590–604) wrote often of this fearsome problem, and many later pontiffs refused to sit in Rome altogether. After the Roman Empire fell, the malaria that made the countryside uninhabitable for popes had the same effects on the invaders who periodically swept down the sunny Italian peninsula. Though some took the *campagna*, none held it very long because their troops, lacking immunity, were devastated by the disease. Visigoths, Vandals, Ostrogoths, and Huns were all afflicted. (It's important to note that malaria was in place elsewhere too, and the Romans found themselves on the receiving end of the disease dynamic. In Scotland, the empire lost more than half of an eighty-thousand-man force to a local strain of malaria.)

The Roman experience in Scotland notwithstanding, sunny Italy was always considered to be the most malarious place in Europe. In the Middle Ages, entire armies were laid low by Italian mosquitoes. In the year 1022, the invading army of Henry II was defeated by a sickness so devastating that a later chronicler declared that it "can in no way be described." French and German-born popes died of malaria in the eleventh and twelfth centuries. Dante wrote about a fever that turned the nails blue "and trembles all."

Elsewhere in the world, writers and religious figures recorded the scourge. Evidence of outbreaks survive from ancient India and Mesopotamia. In ancient China, men traveling to malarious areas were advised to arrange for their wives' remarriage before departing. Many Egyptian mummies have enlarged spleens (a symptom of the disease). Alexander the Great was likely killed by malaria in 323 B.C. Carthage was known to be infected at the time of Christ, and malaria probably helped prevent Genghis Khan from invading Western Europe.

Until the 1890s, no one would know for certain that the fevers, which probably afflicted Lucy and the very first human beings to evolve in Africa, were carried by mosquitoes. But many observers reached conclusions that brought them close to the truth. Time and again, physicians and chroniclers correctly associated dirty, standing water with these illnesses. And they also linked them to travel, armies on the move, and the clash of cultures. Here they were right again. Changing human habits contribute to mosquito-borne epidemics.

Twenty or thirty thousand years ago, when humans lived in small, isolated communities, so did their germs. Malaria, various parasites, and viruses were delivered by mosquitoes, made their attacks, and the body responded with illness and then either death or partial or complete immunity. A certain balance was struck, permitting both the microbes and the human animal to persist and perhaps also to prosper.

The balance began to change when European exploration of remote lands expanded into commerce. The first traders brought their own microbes—really offensive biological weapons—into virgin environments where the bodies of the locals were ill prepared to fight them.

Measles was the most pervasive of these pathogens. Like many infections, measles is far more destructive in adults than in small children and is so contagious that it cannot persist long in a population of less than about a third of a million people. This is because it rages so quickly through a community that in a small population it will run out of human fuel before it can be renewed by births and immigration. Before appropriate vaccines became available, this was seen repeatedly in Inuit villages. The villages would periodically be swept by measles, with adults dying at a much higher rate, leaving children behind. The outbreak would then wane and the virus would disappear until its next introduction, decades later, when a number of people born after the previous epidemic and lacking immunity were available to be infected.

Of course, the people who receive microbes from foreign visitors are not entirely defenseless. Their own local germs are rooted in the landscape and protect them by infecting invaders. Defensive diseases, such as malaria, can generally be maintained in small human populations because they are chronic. The pathogen is constantly circulating in the bloodstreams of local people, who develop a limited sort of immunity that renders their symptoms quite mild. The same malaria that leaves a local resident unscathed often kills the outsider who appears from the wilderness with no established immunity. (This is how the British came to call West Africa the "white man's grave.")

The ebb and flow of defensive and offensive pathogens among human populations was constricted mainly by distance and geography. The damage was limited as long as people stayed at home. But once human beings became capable of mounting armies and navies and conducting long-range military campaigns, the scale of the damage done by encounters with new diseases grew dramatically. Military campaigns were exquisitely sensitive to a kind of warfare by disease, as urban attackers carried their offensive germs among rural people and were met by defensive pathogens lurking in the village landscapes.

A truly tragic story of disease warfare begins with Europe's introduction of offensive germs—most notably smallpox, typhus, measles, and cholera—into the New World. Native peoples have come to regard this event, begun by Christopher Columbus, as a holocaust, and it is impossible to argue the point. Introduced by the first explorers, smallpox burned its way through native populations and by 1520 had killed as many as 80 percent of those in local communities. After thirty years, the indigenous Taino population in Hispaniola—estimates of their numbers range as high as five million—was entirely wiped out by the combination of disease and the predatory practices of the colonists. Were a conquering army to achieve the equivalent level of killing in Europe, it would have had to destroy the entire population of the British Isles, and then some.

European diseases brought similar catastrophe to the Americas. The Mexico Valley population, for example, fell from twenty-five million to one million. This pattern was followed throughout the Americas. By the seventeenth century, British colonists in New England totaled the deaths that illness had visited upon the Indians and declared that God himself had granted the newcomers title to the land.

Unfortunately for the native population, few defensive pathogens such as malaria lay in the American landscape, poised to launch a counterattack on the invaders. However, nature held some deadly ironies for whites who settled in to exploit the land they had taken. After exhausting the local population for slaves to work mines and plantations, Europeans began importing slaves from Africa at the start of the sixteenth century. The ships that brought slaves to the New World—estimates range from twelve to twenty million souls—also carried some passengers never listed on any manifest: mosquitoes, malaria, and yellow fever.

Sometime before 1650, yellow fever traveled to the Americas from Africa, probably aboard a slave ship that also carried its own population of *Aedes aegypti*, known as the yellow fever mosquito. These mosquitoes would have bred in the casks that provided drinking water before the days of steam and distillation machinery. The journey would have taken six to eight weeks, long enough for the virus to cycle two or three times through its human and mosquito hosts. (The incubation period is three to six days both in people and in mosquitoes.)

By the time the vessel reached the Caribbean, many on board would have sickened, the crew more than their human cargo, and as many as a third of those infected might have died. As soon as the stricken vessel docked, the panicking survivors would have fled. Some would have been incubating the virus in their bodies. With their help, yellow fever—called yellow jack by sailors—was loosed on the New World. And it was there to stay, because the virus could invade the forest canopy, where it cycled between the native *Hemagogous spegazzini* mosquitoes and the troops of monkeys that periodically entered their lethal range.

The particular *Aedes* mosquito that came to the Caribbean is *Aedes aegypti*, a beautiful insect marked by artfully arranged black and white lines and spots. Its back bears a delicate lyre-shaped ornamentation.

*Aedes aegypti* is a very sneaky mosquito that prefers to come in low and bite people on the ankles or calves. (This attack mode explains a habit called the Babu Bounce that is practiced by so many secretaries in India. They jiggle their legs up and down almost continually to ward off this little mosquito.) *Aedes aegypti* is very sensitive to movement and will flit away in the middle of probing your skin if you happen to twitch. This nervous habit actually makes her a more effective vector because she's likely to stab you several times during a feeding, depositing pathogens with each plunge of her proboscis.

As a peridomestic mosquito, *Aedes aegypti* depends entirely on people for the small water containers in which she breeds and for the blood that nourishes her eggs. This close relationship with humans—called "anthropophilly"—made the yellow fever mosquito well suited for journeys throughout the world. During a sailing from Africa to the Americas, successive generations of these mosquitoes would have established many new infections. Many historians believe that mosquitoes played a key role in the famous case of the slave ship *Amistad*, which was taken over by its slave "cargo" somewhere at sea. The crew had been struck by yellow fever, to which the slaves were mostly immune.

*Aedes aegypti* mosquitoes are well adapted to travel on the kind of sailing ships used to transport slaves. First, prior to leaving African ports, the ships all took on drinking water in wooden casks. *Aedes aegypti* infestations, as they are now, would have been universal. These mosquitoes lay their eggs at the moist edge of the water remaining in a partially emptied cask. If the casks remain unfilled for a few days, the eggs will mature and be ready to hatch.

The slave ships were supplied with many freshly filled water casks. Adult mosquitoes would emerge a week or so after the casks were filled. A surfeit of blood would then be available to them in the bodies of the masses of people chained in the hold of the ship, which would now be at sea. As the water in the casks was consumed, many more eggs would be laid. No larvae would hatch until the casks were filled once again. But this would happen with each rainstorm, as the crew made sure to replenish supplies. Another cycle of annoyance and disease would follow. Two or three such cycles might have time to occur in the course of the usual voyage. When the ship finally arrived in some American port, it would be shadowed by a cloud of human misery that matched its cargo.

The first *Aedes aegypti* probably made landfall in the center of commercial life in the seventeenth-century Caribbean—Barbados. The island had been uninhabited when the British and then the Dutch settled there in the early 1600s. Sugar plantations were planted and soon began to feed Europe's desire to sweeten two newly popular drinks, tea and coffee. By 1700, per capita sugar consumption in England neared five pounds. It would triple within another seventy years.

High sugar prices and an intense demand fed the constant growth of plantations, which in turn financed the development of a very wealthy colonial settlement. This labor-intensive industry soon became dependent on indentured workers imported from the British Isles and slaves brought from Africa. Labor on Barbados was about as cruel as can be imagined, and the island became synonymous with violence, with the so-called saltwater slaves getting the worst of it.

But the Africans did not get the worst when it came to yellow fever. Many had acquired immunity in childhood and were spared in the epidemics that blazed when the virus and *Aedes aegypti* arrived aboard slave ships. White settlers, who were vulnerable to the virus, suffered greatly. About six thousand were killed by an outbreak that lasted from 1647 to 1650. Another epidemic in the 1690s killed thousands more and prompted a large exodus of whites. Many fled to the North American continent, settling along the coast from New Orleans to South Carolina.

But yellow fever followed them there. The disease scorched virtually every port city in the New World, and ships' crews and soldiers at garrisons were particularly vulnerable. Their officers and doctors responded to outbreaks in contradictory ways. Some, believing the disease was contagious, sought to block infections by quarantining people arriving on ships that had experienced sickness. These quarantines failed when mosquitoes simply ferried the microbe past guards and into the towns. Other authorities feared that the illness resided in the environment and simply fled. Of course, such flights often carried the incubating virus into new sites, thereby extending the outbreak.

During military campaigns in the New World, diseases such as yellow fever and malaria proved disastrous for invading armies. English records indicate that more than half of a twelve-thousand-man force sent to take Cartagena, Colombia, in 1742 was felled by illness. The same fate met British troops later in Cuba. And when the French tried to take back rebellious Haiti in 1802, more losses were inflicted upon them by mosquitoes armed with pathogens than by enemy fighters. Out of a force of 29,000, only 6,000 French soldiers and sailors survived to see Europe again.

The French, who managed to kill more than 150,000 Haitians, were ultimately defeated by the nature of the viral outbreaks. Because yellow fever, like measles, confers immunity on survivors, it will burn out rather quickly as it exhausts the supply of human fuel in a local community. Foreign armies get into trouble because they are fresh wood for the fire. This is how the poorly armed Haitian rebels were able to prevail, while modern French forces fell to infection.

Pathogens also find more fuel by traveling to virgin populations. This is what happens today as the annual winter flu leaves Asia and travels to Europe and North America. In the Caribbean of early colonial times, yellow fever cycled among the islands, often returning to an area it had devastated years before once newcomers lacking immunity had moved in. The virus also accompanied ships to North America, settling in cities as far north as Boston and causing terrible epidemics. One of the worst would strike Philadelphia in 1793. This epidemic came to be regarded as a historic public health disaster and is described in great detail in *Bring Out Your Dead*, a seminal work by J. H. Powell.

Philadelphia in the 1790s was America's busiest port and the home of the federal government. Jefferson, Washington, and Adams resided there much of the time. Benjamin Rush, America's greatest doctor at that time, practiced in the city, but he too would be helpless against the fever.

Records of the epidemic reveal that during a particularly hot and dry summer, ships began arriving from the West Indies, delivering thousands of people fleeing the slave rebellion on the island of Hispaniola. The refugees told of miserable conditions aboard ships rife with illness. By early August, Benjamin Rush had witnessed the hideous death—by fever, hemorrhage, black vomit, and coma—of a local physician's child. More deaths quickly followed, and by the middle of the month, the numbers of sick and dying had begun to cascade. By August 16, it would seem to him that the entire city was sick with yellow fever.

Yellow fever caused panic, and if you consider what it does to the human body, you can understand why. The illness begins with fever, chills, and muscle pain so intense it can feel as if a leg or an arm has been broken in two. Blinding headaches and stabbing pains in the eyes are common. The liver fails. The skin turns yellow. The eyes become red. Blood begins to ooze from the mouth and the nose. Internal hemorrhaging spills blood into the stomach, and this causes a telltale black vomit. Once this occurs, death follows in a matter of days.

The terror associated with yellow fever was amplified by the mystery surrounding its origins and means of transmission. Before the mosquito vector was discovered, debates raged over whether the sickness could reside in a northern city like Philadelphia, or was necessarily imported from the tropics. Those who favored the idea that it was a local problem blamed hot weather, filthy slums, summer hailstorms, even passing meteors for its sudden emergence. On the other side of the argument stood those who insisted that ships from the southern seas brought the disease, and that strict quarantines would prevent it.

Once yellow fever struck a community, doctors were at a loss to explain how it spread locally. Most agreed that one person could not give it to another through direct contact. This theory was suggested by the observation that nurses who cared for the fever's victims were no more likely to fall ill themselves than were other people. As a very impressionable student, I heard Joseph LePrince, then in his old age, describe the heroism of one "Nurse Puleo," whose face and snow-white uniform were regularly drenched by the black vomit of her patients. LePrince was the much honored sanitary engineer who directed the sanitary efforts that permitted the construction of the Panama Canal. Nurse Puleo could do her brave work, as LePrince told us, only because yellow fever is not directly communicable.

In 1793, as ordinary Philadelphians became aware that the disease was in their midst, they anxiously tried to defend themselves. Some built fires in front of their homes, believing the heat and smoke would keep the fever away. Others exploded gunpowder for the same purpose. City fire pumps were used to wash the streets. Newspapers advised people on a number of preventive measures. The most popular was a packet of camphor worn around the neck.

Rush and other doctors became desperate in their struggle to comfort and perhaps cure their patients, whose bloody, fevered, raving deaths outpaced the efforts of the coffin makers and gravediggers. Treatments ranged from bleedings and immersion in cold water to the use of mercury as a purgative. The physicians strained to see some pattern among those who survived, and Rush became convinced that mercury treatment worked best.

This was an illusion. Nothing Rush or the others tried actually stopped the suffering. At the epidemic's worst, people died at a rate of more than a hundred per day, and the demand for nurses far exceeded the supply. Many individuals were abandoned by family members who were too afraid, or too shocked by the suffering they witnessed, to stay near. The deaths of many lonely individuals were discovered only by the stench of their rotting corpses.

Thousands of those who survived the first weeks of the epidemic fled to the countryside. Fearing for the safety of his wife and children, President Washington departed for Mount Vernon. Soon towns and cities along the East Coast imposed quarantines forbidding people from Philadelphia to visit or even pass through. New York sent militiamen into the countryside to enforce the ban. Even members of Washington's cabinet were kept waiting outside the city until they could convince the authorities that they were healthy.

As Philadelphia grew sicker, and the able-bodied departed, the city became quiet by day and ghostly at night. The deaths of mothers and fathers left young children wandering the streets in search of food. Food distribution was disrupted. Many shops closed. Some federal offices stopped functioning; others moved out of the city in order to continue their work. Tents were erected at cemeteries so that gravediggers could rest but return to their duties without delay.

When frost finally arrived, and the adult *Aedes aegypti* began to die, the human death rate dropped. On November 10, George Washington would steal into the empty city, alone on horseback, and discover it habitable once more. With winter, the city began to function as normal. However, some 5,500 out of Philadelphia's 55,000 inhabitants had perished. And the loss of life and fortune would affect Philadelphia for decades to come. Each summer, suspicion would be cast on ships arriving from the West Indies, and people would quietly dread the return of an agent of death that was not clearly identified at the time.

The routine quarantine and rigorous inspection of ships from the Caribbean, piped water supplies, and cold winters would insure that yellow fever would never again rage so relentlessly in a northern city. But throughout the nineteenth century terrible epidemics occurred in the American South, especially in places where *Aedes aegypti* could survive the winter. With its merciless attacks on the lives and fortunes of tens of thousands of people, this mosquito ranked just behind cotton, slavery, and the Civil War as a factor in the region's history.

In the 1800s, New Orleans suffered a dozen major yellow fever outbreaks—each of which killed in excess of 1,000 people—in a span of thirty-five years. Other cities suffered less frequent but nevertheless devastating epidemics. More than 750 of Mobile's 11,500 inhabitants were killed by yellow fever in 1850. Ten percent of the population, 2,800 people in all, died in Norfolk and Portsmouth, Virginia, in 1855. More than 5,000 of the 33,000 residents in Memphis died in 1878.

Families could do nothing more than sit with yellow fever patients, nurse them whenever possible, and witness their suffering. In her 1992 book, *Yellow Fever and the South*, Margaret Holmes quotes a letter in which a father, W. E. George of Memphis, reports the death of his daughter. "Lucille died at ten o'clock Tuesday night, after such suffering as I hope never again to witness. . .the poor girl's screams might be heard for half a square at times and I had to exert my utmost strength to hold her in bed. Jaundice was marked the skin being a bright yellow hue: tongue and lips dark, cracked and blood oozing from the mouth and nose. . .to me the most terrifying feature was the black vomit which I never before witnessed. By Tuesday it was black as ink and would be ejected with terrific force. I had my face and hands spattered but had to stand by and hold her. Well, it is too terrible to write any more about it."

As Holmes reports, the public in other southern cities accepted that yellow fever was a "traveler's" disease that could follow railroad lines and rivers to find new victims. Of course, its precise source remained a terrifying mystery, and the response was therefore blunt and unreasoned. At the height of the Memphis crisis, mobs of armed men met trains bound from Memphis and forced them to keep moving without discharging passengers or freight. People who had left Memphis on foot were barred from entering many towns. Some died for lack of proper food, water, and shelter.

The effect on the city of Memphis's reputation was so profound that after the epidemic some town fathers advocated abandoning the community, leveling its buildings, salting the earth, and reestablishing it elsewhere. Though the plan was never followed, some landowners did sell at cut-rate prices when they believed it would be implemented. Generations later, Memphis would still be regarded as an unhealthy place.

As terrible as the Memphis outbreak became, it could not compare in horror to the mosquito-induced disaster that befell New Orleans in 1853. In an epidemic that at its peak claimed two hundred lives per day, yellow fever killed nine thousand people in the city itself and another eleven thousand in the lower Mississippi Valley. Death overwhelmed New Orleans that summer. Coffins piled up in the delta heat, and the swelling bodies burst their coffins. Desperate city officials burned barrels of tar on street corners and fired off cannons at dawn and dusk in an attempt to "cleanse" the air.

Local doctors believed they had traced the epidemic to a ship that had recently arrived from the Caribbean. But in fact the yellow fever mosquito had probably taken up permanent residence in New Orleans at the time. Winter typically passes in the city without a hard freeze, and *Aedes aegypti* readily survives there. A single traveler from one of many places could have brought the virus into the city. The recipe for an outbreak—one part mosquito vector, one part virus—would be complete. Add a substantial population of nonimmune people, and a disaster was inevitable.

For much of the century, yellow fever shaped migration and economic development in New Orleans and throughout the South. When outbreaks emptied cities, those who fled were slow to return, and the rest of the nation stayed away entirely. Investment, immigration, and trade were all limited by disease and by rumors of disease. The more superstitious read something ominous into the plague of yellow fever. Many of the pious saw it as a by-product of sin.

Reports on the pall of sickness that hung over the South reinforced the prejudices of Northerners, many of whom came to regard the rest of the country as backward and dangerous. In one long sentence, a medical historian named F. H. Garrison summed up the region's inhabitants. They were "pale, gaunt, haggard, attenuated, narrow-chested, spindle-shanked, sharp featured, lantern-jawed, lank-haired, anxious-eyed with care furrowed brow, of pasty, sallow, bilious or dyspeptic complexion or serious concentrated, careworn expression and languid or irritable mien."

Sheldon Watts notes in his book *Epidemics and History* that the persistent presence of yellow fever also contributed to the rise of a very dangerous social environment for blacks and foreigners. Many southern physicians contributed to the pro-slavery mindset by declaring that blacks were immune to yellow fever and thus were members of a servile race better suited to work the plantations. Dr. S. A. Cartwright of New Orleans concluded that whites were vulnerable to the disease because "nature scorns to see the aristocracy of the white skin. . .reduced to drudgery work."

Far more direct was Henry Rose Carter, M.D., who was the chief medical officer for the state of Mississippi after the Civil War. Carter definitively declared that yellow fever was "imported by Negroes" who then spread it around the countryside. At the turn of the century, Carter would continue to espouse this idea, even as a member of the U.S. commission, headed by Walter Reed, that studied yellow fever in Cuba.

These "medical opinions" contributed substantially to the racism that dominated life in the South. And when outbreaks occurred, whites relied on such shaky science to justify mob vengeance. The Ku Klux Klan set up roadblocks and quarantines to prevent the movements of blacks and immigrants. In one of the few lynchings that was actually reported in local newspapers, five immigrant Italians in Louisiana, who were blamed for spreading yellow fever after socializing with blacks, were hunted down and then hanged by a mob.

Nothing that a mosquito could deliver to a nineteenth-century American rivaled yellow fever for its killing power. Nevertheless, malaria was also a serious problem, and its effects lasted much longer. Yellow fever's survivors resumed more or less normal lives in a few months and won the distinctly valuable prize of lifelong immunity from the ravages of the virus. After an acute illness, malaria could linger to affect a person for the better part of a year and conferred immunity that might only last a short time. Many people seemed to get it over and over again.

The listlessness associated with malaria in the Midwest and the South kept investment and immigration out of the region. Mark Twain noted the fevers that bloomed along the Mississippi. And in 1850, the frontier physician Daniel Drake published a long account of the problem, describing epidemics that raged throughout the "interior of North America." Lincolnesque in appearance, Drake was a professor of medicine and one of the most famous doctors of his time. Though his education was minimal, he was a scholar and a healer whom Lincoln himself once contacted about an illness. (Drake refused to treat Lincoln via the mails, and the two never actually met.)

In his treatise, Drake recounts an investigation that he conducted as he traveled over thirty thousand miles, making inspections of entire regions from the Gulf of Mexico to the Canadian border. He noted where people lived, what they ate, the conditions of their homes, even their wardrobes and exercise habits. He described malaria epidemics where hundreds died of fevers and cataloged outbreaks that occurred in military garrisons.

Nothing in Drake's account is more disturbing than his descriptions of the way patients were treated for their illnesses. In many cases the cure must have been much worse than the disease itself, for it involved potions that induced vomiting and diarrhea as well as bleeding and even burning the skin. It is no wonder that many patients avoided doctors altogether.

Given how wrongheaded medicine was at the time, it's all the more remarkable that Drake was able to document a great many certain facts about malaria outbreaks. He noted that the disease is more prevalent in warm, southern climes and that it follows a cycle that sees infections peak in early autumn. He also linked the disease to water—rivers, creeks, ponds, and marshes. Through careful reasoning, he concluded that heat and moisture are not enough to produce malaria. "A deleterious agent, diffused in the atmosphere" must be involved, he wrote. Some living thing is ultimately responsible, observed Drake. But he stopped just short of naming it.

As troublesome as malaria was to the growth of the American West, it was an even more significant agent of history in Africa and Asia. A powerfully effective defensive pathogen, malaria was well established throughout the tropics when Europe began to seek colonial outposts in the 1400s. In many of these places, the parasites and local peoples had developed an exquisitely balanced relationship. As the pathogens became endemic, immunity became almost universal. Except for women in their first pregnancy, the only adults that could possibly be killed by this pathogen would be foreigners. In this way, the microbes were like a defensive wall.

Portuguese traders and explorers of the late 1400s and early 1500s were probably the first foreigners to run headfirst into the wall of malaria and yellow fever that protected Africa. For the next three centuries, European powers struggled to establish colonies on the continent but were often defeated by these diseases. The most enduring story of this deadly dynamic involves the fateful expeditions along the Niger River that were mounted by Scottish explorer Mungo Park.

At the time of his first journey to Africa in 1795, Park entered a land that Europeans considered forbidding. Many of his predecessors had died of disease or simply disappeared in the wilderness. With the help of an African who had learned English as a slave in Jamaica but returned to his homeland when he was freed, Park managed to march four hundred miles into the interior from the shore of the Ivory Coast. He failed to accomplish his goal, which was to discover the headwaters of the Niger River. But upon his return to England he promptly published a book and established himself as the archetype of the intrepid British explorer.

Only Park, four men, and a simple journal survived from Park's second expedition, which began in 1805. This time, no Africans accompanied him, only whites recruited in Europe and at coastal trading ports in Africa. From the very first week, the long column of men and pack animals was raided by thieves and ravaged by disease. Between accounts of lions and crocodile attacks and run-ins with local tribal leaders, Park hinted at the sickness that hovered around his troop.

June 8—"The carpenter was unable to sit upright and frequently threw himself from the ass [donkey] wishing to be left to die. Made two of the soldiers carry him by force."

June 9—"The soldier who had been left to take care of the

sick man, returned and informed us that he died...five of the soldiers...complained much of headache and uneasiness at stomach."

June 11—"Twelve of the soldiers sick."

June 15—"Men still very sickly, and some of them slightly delirious."

June 18—"Lt. Martyn, the sergeant, the corporal and half the soldiers sick of the fever."

Park managed to recruit locals to help carry provisions through the grasslands and forest. At the same time, word of an expedition of weak, sickly, and well-provisioned whites spread through the countryside. As bandits descended on them, Park and his men had all they could do to defend themselves with their firearms.

By July, Park's journal was filled with his efforts to help his men with doses of boiled fever bark and as much fresh meat and milk as he could procure in the villages they entered. Many of the men who were too sick to ride begged to be left at various points along the trail. Park complied, knowing they would likely die where they lay.

July 6—"All the people sick, or in a state of great debility, except one." A few days later he added, "continued sickness and deaths."

Park recorded deaths almost every day, and yet he pressed on. Eventually the men were so weak that they could not build rafts for crossing streams or even replace a horse's load once it had fallen off. Jackals tracked the party, and listless men awoke in the night to find them nipping at their feet.

After more than a hundred days in which his party grew ever more desperate, Mungo Park accepted that he would not reach his goal and turned back. The last death recorded in his journal came on October 28. Alexander Anderson, his wife's brother, died after a full four months of illness. This tragedy offered Park a chance to make a journal entry that, given all the death he had seen, is appalling. "No event which took place ever threw the smallest gloom over my mind," he wrote of his entire expedition, "until I laid Mr. Anderson in the grave."

Mungo Park returned to England with just four of the forty-four men he had brought into the interior. One of those was judged to be mad and reportedly never recovered his sanity.

Park would return to Africa one last time, where he would drown in the very same river he followed in 1806. Although an African who had accompanied Park reported on the explorer's death, his family refused to accept it. Mrs. Park died still expecting word that he had emerged from the wilderness. Park's son Thomas was so certain that his father was alive that he went to the Ivory Coast himself to search. He too was never heard from again.

The mosquitoes that killed Alexander Anderson and most of Park's men made imperial Europe's adventures in Africa costly. In Park's time, Britain ruled Sierra Leone and the Cape Colony in what is now South Africa, but would move very slowly in the rest of the continent. Though political power might be seized, much of Africa could not be inhabited comfortably by Europeans who lacked immunity to malaria and yellow fever.

Between 1819 and 1836, almost half of the British soldiers sent to Sierra Leone died. In roughly the same period, the French suffered losses of 16 percent in Senegal. In Algeria on the coast, France found a slightly more hospitable environment, but the death rate from disease was still high. Garrisons were perpetually undermanned, and the expense of keeping forces at full strength was high. Even when territory was deemed safe by authorities, settlers stayed away out of fear.

More than a century would be required for Europe, mainly Britain, Germany, and France, to fully establish itself in Africa. To aid the cause, the British government established schools of tropical medicine both in Liverpool and in London. These schools devised both medical treatments and public health strategies to aid the colonial effort. One major "contribution" to social policy made by these schools was a system of segregated housing, which was used throughout Africa, to keep Europeans and blacks separated. A century later it would still be in evidence, as apartheid in South Africa.

But even at the height of Europe's influence, white settlement would be concentrated at higher elevations and in a few port cities and mining centers where strictly enforced sanitary measures could improve the quality of life. One out of every ten lower-class Englishmen and one out of twenty of their upper-class countrymen are said to have died in the Bight of Benin each year, mainly of malaria. A well-worn rhyme recalled the danger:

*Beware beware, the Bight of Benin.*
*One comes out where fifty went in.*

As a result of malaria, European dominion over many sections of the African continent was never fully established. And while geopolitics, logistics, and military resistance helped exclude the Europeans, the *anopheline* mosquito may have been the most important factor.

Since no one understood the dynamics of malaria, whites and

black Africans were challenged to explain the suffering of the Europeans. Each approached the problem with a certain bias. Many whites believed their race possessed superior but delicate brains that were vulnerable to the sun. The pith helmet, with its wide brim, was invented to protect the head from this threat. Many Africans however concluded that the whites were lesser creatures. An African song, published in a nineteenth-century history, reflects the way many locals saw the intruders.

*The poor white man, faint and weary*
*Came to sit under our tree.*
*He has no mother to bring him milk;*
*No wife to grind his corn.*
*Let us pity the white man.*

Conditions were not much better for Europeans in India, Asia, or the South Seas. The British in India were constantly dealing with mosquito-borne diseases, especially malaria. In the early 1800s, the non-combat death rate for the empire's soldiers in India was twice the rate for those in Great Britain. Many of these deaths were caused by malaria.

But with its huge trading potential, India was vital to the Crown, and greater effort was expended to make certain areas habitable for Europeans. An Indian Medical Service was established and filled with doctors and nurses whose mission was to bring English settlers and English quality of health. Along with tending the sick, the service also strove to improve overall health conditions.

Eventually British authorities in India would implement the public health strategies advocated in Europe by so-called sanitarians. Just like the ancient Romans, the burghers of nineteenth-century Europe had come to associate clean running water, working sewers, and tidy streets with good health. In the 1830s and 1840s, these ideas contributed to a social movement advocating public sanitation projects. The sanitarian crusade was imbued with a high Protestant sense of morals. Its soldiers firmly believed that sin, in the form of sloth and decadence, was at the heart of much disease.

Edwin Chadwick, of the English Board of General Health, was at the head of the sanitarian parade. Working from the miasma concept—which linked disease to foul airs—Chadwick published a book in 1842 that advocated the construction of sewers and water systems under streets to rid cities of disease-laden stench. Beginning in the 1850s, water and sewer systems were under construction throughout Great Britain, and public health began to improve. Similar campaigns were started in France, Germany, and much of the rest of Europe, with the same beneficial results.

Many Americans also embraced the sanitation doctrine, though they were slow to understand the need to protect drinking water. In New York, Boston, Washington, D.C., and many other places, low-lying areas received drainage systems, sewers were built, and roads were paved. In 1880, Memphis built a sanitation system designed by New York's George Waring, and the city would never again suffer a major yellow fever epidemic. Across America great effort was made to remove trash, including dead animals, from city streets, and food inspections began.

Though they were often associated with unpleasant places—sewers, swamps, waste pits—that the sanitarians attacked, mosquitoes and their biting behavior were not much mentioned by the crusaders of cleanliness. These crusaders linked illness to foul-smelling filth—not insects—and were convinced that health arrived because stench disappeared.

Decades would pass before sanitarians learned why their efforts succeeded in reducing malaria and yellow fever. It was not cleanliness or even godliness that saved humanity but rather the suppression of the tiny mosquito. This fact would come to the fore very gradually, and it would be propelled by the discovery and exploration of a tiny world that had always existed, just beyond view.

# 5

# VECTOR

The compound microscope, with high-quality lenses that produce clear images, was a product of the 1820s. At the time of their development, these instruments were so expensive and scarce that their use was limited to the most rarefied research institutes. The first major breakthrough born of this technology was Rudolf Virchow's observation that human beings are made of cells.

Initially, Virchow's discovery was met with a blend of indifference and even hostility from the broader medical community. But among those who had access to microscopy, he immediately set off a race to discovery. Louis Pasteur and Robert Koch were his leading competitors. Both isolated bacteria that caused many biological processes, including human disease.

For several years the medical establishment resisted the findings of Koch and the others, clinging to a centuries-old tradition, which held that while illness had a great many causes—sloth, miasmas, immorality—tiny living creatures were certainly not one of them. But gradually the tide of evidence viewed through the microscope became so great that it overwhelmed the so-called great tradition of medicine. The publication of Pasteur's *Germ Theory of Disease* in 1880—just two years after the Memphis yellow fever disaster—heralded the wide acceptance of the new biology. It began a period of amazing advances, every bit as exciting as the space race of the 1960s or modern genome research today.

Far more important at that time than any single discovery was the dramatic change in medicine's overall view of the body and illness. Physicians began to see the human organism as a place where battles over health and disease were waged at a microscopic level. Scientists raced to find the pathogens that invaded the body and the means they used to gain a foothold. An era of great men and great advances began. The competition to make these discoveries was fueled by personal ego, nationalist pride, and, in some cases, the hope for financial gain. None of the breakthroughs received more attention than those that connected the mosquito to the transmission of disease. The first of these involved a comparatively rare illness that could produce a remarkably grotesque set of symptoms.

## FILARIASIS

In the nineteenth century, tropical disease was a major concern for the British government, as its empire reached around the globe. In Africa, Asia, and India, colonial officers and their families were terrified by a host of illnesses, which claimed lives, slowed the march of conquest, and exacted an enormous financial price. Few of these diseases caused disfigurement and suffering in a way that even approached elephantiasis, the most extreme manifestation of infection by the filaria worm. No one who confronts a patient suffering from this disease ever forgets it.

Elephantiasis is one of the most stigmatizing diseases in the world. It occurs when the flow of lymphatic fluid is blocked and parts of the body—frequently the legs or genitals—swell enormously. Under the pressure of the swelling, the skin stretches until it becomes papery thin. Fluid collects, often forming thick sheaths that flow from a person's hips down his legs.

In males afflicted with elephantiasis, the scrotum can swell to the size of a medicine ball. In women, the labia may become almost as large. If the genitals are spared, bags of fluid may form on an arm, or leg, surrounding the limb like so many rubber inner tubes. Feet and ankles are especially vulnerable, and can come to look quite elephantine. In some cases, people develop acute infections of their skin. The disease is almost never fatal, but in extremely rare instances it can be a factor in death. For example, due to secondary infection an arm or leg can become gangrenous, and that infection can kill.

Not surprisingly, elephantiasis made a profound impression on Dr. Patrick Manson, who in the 1870s was a British medical officer in Taiwan, which was then a part of China and was called Formosa. As personal physician to hundreds of Europeans and locals, Manson saw dozens of cases. No drug was available to treat elephantiasis, and surgery, including castration, was often the only option for relieving symptoms. Manson drained and surgically removed the bloated flesh caused by this disease, and he became obsessed with discovering just how such a dramatic illness was acquired, and how it might be avoided in the future.

In 1875, when Manson visited London on a break from his service in China, he bought a compound microscope. After a visit to the family home in Scotland, thirty-one-year-old Manson boarded a ship for the long journey back to China. For the next nine years he would practice medicine under the authority of the British Customs Service and would pursue the minute animal that causes elephantiasis.

Though autopsies were a crime in China, Manson secretly opened the bodies of several people who had died while suffering from elephantiasis. In one of these cadavers, he found thread-like female worms, about as long as a person's arm, which lay tightly curled with their much smaller mates within the body's lymph nodes. The disease Manson saw in his patients was caused by the victim's immune system reacting to the presence of the copulating worms.

Manson viewed nature's efficiency and ingenuity with a genuine awe that he would later communicate to generations of students. Noting that "nothing walks with aimless feet," he understood that the filarial worm could not spend its entire life cycle locked within a human body. Its offspring must somehow be transported from the bloodstream to the outside environment, where it could then find its way into another human host. The most obvious suspect was the one creature whose life seemed most intimately tied to human blood, a mosquito.

In training his eye on mosquitoes as an agent in human disease, Manson was not posing an original theory. More than twenty years before, Daniel Beauperthuy in Venezuela and Josiah Nott in Alabama had accused the mosquito of transmitting both malaria and yellow fever. Nott and Beauperthuy had themselves followed a trail that was marked more than a century earlier by an Italian physician named Giovanni Lancisi. Having noticed that malaria receded when swamps were drained and mosquitoes departed, Lancisi postulated that the insects carried some sort of swamp poison to humans.

All this early conjecture was based on circumstantial evidence and logic. Not one of the theorists had actually demonstrated that mosquitoes might bear a substance that could make a person ill. This made it easy for critics to overwhelm the mosquito theory. The idea was regarded as so outlandish that Beauperthuy became the object of ridicule among scientists in Europe, and some even considered him insane.

To establish a direct connection between mosquitoes and human disease, Manson needed to find his filarial worm in the blood-filled gut of an insect. For this he turned to his gardener, a Chinese man named Hinlo, whose blood teemed with the tiny filarial parasites, each about as wide as a red blood cell but many times longer. As with many other infected people, Hinlo manifested no symptoms of this infection. He agreed to sleep in the screened room that Manson dubbed the Mosquito House, and into which the doctor released mosquitoes that later fed on his sleeping gardener. The mosquitoes were recaptured the following morning, and Manson dissected them during the next five days. He would later write: "I found that the haematozoon [Manson's name for an early larval developmental stage of the parasite] which entered the mosquito a simple structureless animal and left it, after passing through a series of highly interesting metamorphoses, much increased in size, possessing an alimentary canal, and being otherwise suited for an independent existence."

Manson had discovered that the filarial worm lived and developed in both man and mosquito. Although he thought that these worms completed their development in mosquitoes by five days after feeding, the actual developmental period is more than twice this long. Manson's worms had not yet matured. Manson made another major mistake. Believing, as physicians of that day generally did, that the mosquito fed only once in its lifetime, he theorized that the parasites infected people only when they escaped into drinking water from the disintegrating body of an infected mosquito.

Such mistakes are minor, however, when compared with Manson's monumental accomplishment. He had firmly established that mosquitoes performed an essential role in generating human disease, and advanced research on myriad tropical illnesses, most especially malaria, in the direction that would soon reveal the mosquito's role in producing explosive outbreaks of disease. Other researchers would chart the entire complex cycle of the filarial worm, both in the mosquito and the human body. Today we know that filariae matures after a thirty-day incubation period, and years can pass before elephantiasis is evident.

Manson's choice of the correct mosquito was guided by good fortune and the site in which he worked. Much of his work was conducted in Kaohsiung in southern Formosa, a built-up city that, even today, harbors huge populations of the southern subspecies of the common house mosquito, *Culex pipiens quinquefasciatus*, which entomologists refer to as "quinques." Like *Aedes aegypti, Culex pipiens* originated in Africa and was distributed around the world by the early sailing ships. These mosquitoes, however, were much more abundant in the filthy bilges of these ships than in the much cleaner water casks. *Aedes aegypti* die in the sludge in which *Culex pipiens* larvae thrive. In adulthood the quinque feeds mainly on human blood and on sugar, when available. They are drawn to foul-smelling water for oviposition. Kaohsiung offered much ideal habitat for quinques.

The microfilaria that a mosquito might swallow with its meal of human blood is the first in a series of developmental stages deriving from a large worm residing in a lymph node. Once within the mosquito's gut, the larvae bore into the insect's flight muscles and develop there through two more larval stages before they travel to the insect's head. The third-stage larvae wait within the double-walled sheath that covers the mosquito's feeding stylets, poised to break out through the sheath's inner wall when it is retracted to permit the feeding apparatus to penetrate the skin of a luckless person.

The strain of retraction causes the mosquito's own blood to flow out onto the skin around its bundle of slithering stylets, thereby forming a pool in which the now-infectious larva swims. The feeding mosquito soon withdraws its mouthparts. The worm then bores its way through the wound into the skin of this, its next host, and ultimately finds its way into the nearest lymph node. Michael Lavoipierre, of the University of California at Davis, designed a clever experiment to demonstrate that their larvae need to follow a trail of blood into the wound made by a mosquito's bite. He did this by allowing a mosquito to begin feeding on a gerbil, and then severing the mouthparts before the proboscis was withdrawn. With the wound plugged by the mosquito's feeding stylets, the larval worm could not enter the animal's body and cause an infection.

The worms that cause this variety of human lymphatic filariasis mature solely in people. After a male and female worm find themselves in a lymph node, the cycle can be completed. The fertilized eggs hatch within the mother worm's uterus, producing microfilariae that are shed into the lymphatic circulation, pass into the human host's abdominal cavity, and from there drain into the bloodstream.

As with many infectious diseases, it is the victim's own immune response to the presence of the pathogen that produces symptoms. This interplay of the parasite and the immune system is complex and variable. For years it was known that adults may acquire a kind of chronic pneumonia after visiting a region in which filariasis is endemic. These symptoms result from an inflammatory reaction that traps the microfilariae in a person's lungs; no worms are evident in the bloodstream. This condition is known as tropical eosinophila. Some twenty thousand American servicemen developed this disorder during World War II, but not one developed elephantiasis.

Among the people who live where filariasis is rampant, the extreme lymphatic response that produces elephantiasis is relatively uncommon. Indeed, in one survey in Calcutta, microfilariae could be seen in 20 percent of the general population. Some of the infected people experienced mild lymphatic obstruction and only a few the more extreme manifestation of elephantiasis.

In experiments we conducted with gerbils, Faye Schrater and I found that an animal's mother played an important part in its own response to filarial infection. Animals born to an infected mother were themselves more susceptible to infection, and microfilariae were far more likely to appear in their blood than in that of gerbils whose mothers were not infected. In utero exposure to these worms seemed to "tolerize" the resulting animal, causing it to recognize these worms as "self" and modifying their immune response.

Microfilariae were filtered out in the bodies of animals born without this "immune tolerance" before they appeared in their circulating blood, just as they were in the bodies of the American soldiers who became infected during WWII. It may be that elephantiasis develops in people whose mothers were infected by these worms but who were, themselves, infected relatively late in life, after their tolerance partially dissipated.

Our pet dogs acquire a similar kind of infection, another filarial worm known as dog heartworm (*Dirofilaria immitis* and *repens*). In many communities in North America and in southern Europe, a third of unprotected animals may be infected. These huge male and female adult worms lie coiled within the chambers of the animals' hearts, not in the lymph nodes, as in the case of human lymphatic filariasis. Cardiac function may be affected severely, virtually crippling the animal. Microfilariae are released from the uterus of the female. Certain mosquitoes serve as vectors of this disease, transmitting it from one canine to another.

Very infrequently these worms, which normally infect dogs, may infect a person. But the developing worm then lodges in a lung and never makes its way to the heart. A node forms there that may cause a false diagnosis of lung cancer when viewed by X ray. Some people have experienced unnecessary surgery and lung resection for this reason.

Dogs are routinely protected against heartworm by the administration of certain drugs all through the summer mosquito season. Until recently, a daily dose of diethylcarbamazine (DEC) was required, but the drug carried some risk. It destroyed the worms, causing them to release their body contents into the animal's bloodstream. This could cause shock and even death in a previously infected animal. Fortunately a new drug, ivermectin, was developed. It sterilizes the worms and stops their development without risk to the dog. It has become the prophylactic drug of choice.

The same drugs that are used to treat dog heartworm also are used to treat human filarial infection. Ivermectin is one such drug, eliminating infection before extensive damage is done and frequently serving to reduce the force of transmission in a community. The gross signs of elephantiasis, however, still require surgical correction.

River blindness, or onchocerciasis, caused by *Onchocerca volvulus*, is another human filarial infection that is treated effectively by ivermectin. This worm, however, is transmitted by black flies, small mosquitolike bloodsuckers that breed in fast-flowing streams. The intertwined adult worms cause a cherry-size nodule to form around them beneath the infected person's skin. Their microfilariae migrate just below the outermost layer of the skin and through the surface of the eye. The resulting inflammatory reaction may be intense, causing intolerable itching and scarring that results in progressive loss of vision. Although some parts of Africa and Latin America had been rendered virtually unlivable because of these worms, a far-reaching program led by the World Health Organization has resulted in vast improvements due to regular applications of insecticide and ivermectin. The medicine is donated by an American drug company, in a program that is reportedly paid for largely by the sale of drugs for the protection of dogs in the United States.

Today's medical treatments for the various filariases are extremely effective, but the rapid expansion of communities in regions in which these infections are endemic exposes far more people to risk of these diseases—hundreds of millions of people—than were exposed in Manson's time.

## MALARIA

In the late 1870s and early 1880s, Manson's pursuit of the filarial worm was just one of many tropical disease investigations being conducted at a fevered pace around the globe. Malaria was of greatest interest, and scientists in France, Germany, Italy, the United States, and Great Britain competed with nationalist fervor to uncover its secrets. The struggle was no less intense than the race to discover the New World had been, and it foreshadowed future scientific contests over the polio vaccine and the atom.

Widespread and debilitating, malaria was the most common of the tropical diseases. Virtually no one who visited parts of Africa, Asia, or Latin America escaped infection. Malaria's symptoms are unmistakable. Seven to fourteen days after an infecting bite, malaria begins with a chill that spreads like Jack Frost on a windowpane. The victim's skin becomes pallid, and shaking begins. Then the cold plunges to the body's core and deep tremors—called rigors—come in waves. Though weakened by the disease, a man who feels the ice of malaria will shake with a violence that can propel an iron bed across the floor. Completely out of the victim's conscious control, the spasms are the body's vain attempt to generate heat. Of course, the man, woman, or child who lies in the parasite's grip soon comes to understand that fever will inevitably follow the chill, and bring the next stage of suffering.

The fever from malaria can reach 106 degrees Fahrenheit and brings a sweat that is so profuse that doctors can almost see the droplets forming on a patient's forehead. They certainly see the rivulets that run down the temples and onto the pillow. The bedclothes and sheets of a malaria victim become soaked with salty perspiration that is the by-product of the body's valiant effort to cool and protect vital organs.

Death often comes to those who are physically weakened by age or previous illness. The dying languish while their red blood cells, made sticky by the parasite, form a plaque that clogs their vascular systems and starves the brain. Most follow the same pathway: lethargy, followed by delirium, then coma.

Even those who do survive their initial disease episode cannot consider themselves immediately cured of malaria. In adults and children the progeny of the original parasites remain in the body for months or years. For the most part they are held at bay by the immune system. But a decline in immune function—due to stress, fatigue, malnutrition, or another illness—can allow the disease to return in all its virulence.

Partisans still argue over who first reached certain milestones in the investigation of malaria. But there is no doubt that a French Army doctor named Charles Laveran was the first to see malaria parasites in the blood of an infected person. He made the discovery while working as a doctor for the French Foreign Legion outpost in Constantine, Algeria. It was out of colonial concern, not scientific curiosity, that the French government sent Laveran to Algeria in 1878. Military officials were desperate to stop the epidemics that plagued their outposts. The sick and the dead among the Foreign Legion had to be replaced, and the cost in both cash and morale was astronomical.

There were few better places to study malaria. The disease was endemic in Constantine, and the soldiers of the legion were wholly susceptible. Laveran wrote that upon his arrival he immediately "had the opportunity of making autopsies on subjects who had died from pernicious attacks."

On November 6, 1880, the thirty-five-year-old Laveran drew blood from a soldier who was wracked by fever. He placed a drop under his microscope and saw that it was alive with little moving animals. Laveran was sure that this protozoan, not a bacillus previously found in the soil by an Italian-based group, was the cause of malaria.

Laveran's work eventually was confirmed in major laboratories, and he alone was credited with the discovery that eventually earned him a Nobel Prize. But it fell to a British doctor to demonstrate how the malaria parasite entered the human bloodstream.

He was an unlikely, if not reluctant, scientist. Eldest of a British general's ten children, Ronald Ross was born in the foothills of the Himalayas in 1857 and educated in England to pursue an artist's life. He loved literature and music. A true polymath, young Ross showed a gift for mathematics and was so enthralled by engineering that he attempted the folly of designing a perpetual motion device. Medicine became his life's work only at the insistence of his father. At age twenty-seven, having become a member of the Royal College of Surgeons, he was posted to Bangalore by the Indian Medical Service.

Many of the British physicians who went to India at that time were drawn by the life of ease the service promised. The demands of the job required just a few hours each day. This left much time for golf, shooting, and fishing. In the first years of his posting, Ross indulged in these pursuits and more. He wrote poetry and began a romance novel, which was eventually published in London. But even with this success he was bored and yearned for achievement.

In 1894, he returned to London and experienced the excitement generated by medicine's shift toward Pasteur's germ theory.

In London, Ross sought out Patrick Manson. An often-told story has the two men walking from Manson's home on Queen Anne's Street down Oxford Street to Seaman's Hospital, where Manson had taken a position on his return from China. At fifty, Manson was a full thirteen years older than Ross. But they had much in common: postings abroad, an interest in tropical disease, scientific passion, Scottish blood. And when Manson said that he agreed with the theory, then circulating, that mosquitoes might carry malaria, Ross replied that he suspected the same thing.

At the hospital, it took Manson just a few minutes to show Ross a few of Laveran's creatures in a blood stain. (Ross, less skilled at microscopy, had spent hundreds of hours looking for them in vain while in India.) Manson, and many others, had seen the malaria parasite develop in the blood, transforming itself from a quiet, crescent-shaped structure into a more motile, obviously living organism.

In the wild, Manson speculated, mosquitoes were what he called "nurses." They acquired the parasite from people in a blood meal and then protected it as it developed into the motile form. He was right to believe that the malaria parasite was essentially adapted to live in the human body. But he mistakenly presumed that it was transmitted in drinking water, the same mistake he had made in regard to the filarial worm.

When Manson published his theory of the mosquito as nurse, he was challenged in writing by Amico Bignami, a leading malariologist in Rome, who was certain that the parasite could not be dependent on the mosquito for its development. Bignami was wrong about this, but he was right in one conjecture he made in the same paper. He argued that the mosquito took up the parasite and transmitted it by way of its bite.

Bignami's idea was not as bold as it may have seemed. Certainly he would have been aware of American Theobald Smith's demonstration that a blood-feeding tick was transmitting a malaria-like infection among cattle in the American west. Smith, who would later help found Harvard's School of Public Health and the Rockefeller University, conducted three years of experiments that led to his inarguable conclusions, well before Manson and Ross took their walk down Oxford Street. His 1893 observation was the first to demonstrate that a blood-feeding creature could acquire and transmit a pathogen via its bite. This first paper, which also laid out the blueprint for the elimination of this disease from North America, was an amazing advance.

It is easy to imagine the excitement that Ronald Ross felt as Manson—a true giant of medical science—brought him into the fraternity of malariologists. After Manson showed him the parasite through a microscope, he then shared what he knew and what he believed about its development and transmission. At last Ross had not only an outlet for his ambition and intellect but also a mentor who provided direction. He returned to India determined to prove Manson's "Grand Induction" by observing malaria's growth and development in mosquitoes.

It was a tall order for a man who hadn't been able to find the parasite on his own, didn't know one species of mosquito from another, and had never dissected one. But he was so eager that he examined the blood of every passenger on the ship to India, and hunted for malaria in the ship's ports of call. In the middle of the ocean, when a flying fish landed on the deck, he examined it too for parasites.

Once in India, Ross kept Manson apprised of his work in a steady stream of letters that were later published. They reveal a man who was constantly vexed, by everything from mosquitoes that wouldn't bite to the weather.

"The heat is awful, and the wind like the blast of an engine-furnace," he reports in his first letter after his return to India. "I am wild with excitement; but these mosquitoes. They will not bite!"

Ross killed an entire collection of mosquito larvae he intended to raise by leaving them exposed to sunlight. His duties as a physician drained time from his research. And he was frustrated by local doctors who wouldn't give him access to their malaria-stricken patients.

He recounts how even when he did get human subjects, they and the mosquitoes proved most unreliable. "When with much difficulty I had got two [human] cases out of the bazaar, all my mosquitoes died again; the next day, when I had got a fresh crop of mosquitoes, my two cases ran away, because I pricked their fingers, in spite of my giving them a rupee a prick!"

Manson was an ever-encouraging teacher, colleague, and father figure. He helped get Ross's papers published in scientific journals and pressured the government to support his protégé's work. And he encouraged Ross to persevere in a way that would surely motivate a younger man blessed with equal amounts of ability and ego.

"Peg away; you are doing most valuable work and I firmly believe that a light will dawn on you soon such as will astonish you and the rest of mankind and be a complete reward for all your efforts."

Over time, Ross and Manson fell into a comfortable dialogue. They referred to the malaria protozoan as "the beast" or "the little brute." Manson mused that he was "born too soon" because the future held such wonder. He would also gossip in letters, writing for example about a fellow British scientist who "has made such an ass of himself" and was being ridiculed in

America. For his part, Ross was eager to impress Manson with his efforts, reporting that constant work "has reduced my brain to mush." His letters were full of complaints, frustration, apologies for delays, and promises to press on.

For three years Ross struggled. His work was interrupted by a cholera epidemic in Bangalore, which he tried to combat with a sanitation program that got little support from his fellow health officers. Millions died as Ross's recommendations for controlling sewage were ignored. (Ross was so convinced of the need to scrub cities clean that he even wrote a poem on the theme titled "Great Is Sanitation.")

Returned to his research, Ross was frustrated in his attempts to find the malaria parasite in the two kinds of mosquitoes he dissected—a gray one (probably *Culex pipiens*, the common house mosquito) and a brindled one (probably *Aedes aegypti*, the yellow fever mosquito). He spent long days doing nothing but collecting mosquitoes, coaxing them to feed on malarial blood, and dissecting them to see if he discovered the developing parasites. In one mosquito after another, his painstaking surgeries revealed nothing.

Finally one of Ross's Indian assistants nurtured the larvae of ten mosquitoes that developed into "big brown fellows," a type Ross had never seen before. These were allowed to feed on patients suffering from malaria and were then returned to individual test tubes.

With a puff of smoke Ross killed two of the freshly fed mosquitoes and examined them. He saw no parasites. By the next morning, two of the remaining mosquitoes had died. Ross killed and then tried to dissect another two, but clumsily ruined them. Just four remained. Number seven was dissected and Ross saw nothing. Number eight died. Four days after the mosquitoes were fed, on August 20, 1897, Ross decided to examine the remaining two insects.

The story of what happened next was later refined, embellished, and even turned into a bit of a myth by Ross himself. (He would add to the tension of the story by saying that he slowly examined various parts of the mosquito before turning to the gut. In fact, he went straight to it, as mosquito anatomy dictated.) By the most accurate account, Ross opened the gut of the ninth mosquito and immediately saw dark, pigmented cells that were not part of the creature herself. Ross later recalled that he was so familiar with the inside of a mosquito that "these bodies struck me at once." Their texture suggested that they were identical to the malaria parasites found in human blood. A day later Ross found the same structures—only larger—in the stomach of the remaining mosquito. This confirmed that he was dealing with a living, growing organism.

It fell to others to discover that certain mosquitoes—notably various anophelines—acquire malaria while others do not. (Ross left India still confused on this matter.) And the details of the parasite's life cycle would also come from various sources. Working with bird malaria, W. G. MacCallum, a Canadian studying at Johns Hopkins University, was the first to observe that male protozoa joined females to create zygotes, an entirely new form. An Italian team led by Giovanni Battista Grassi showed that human malaria followed the same process.

Inspired in part by MacCallum, Ross began to work out the final mystery of malaria—how it was transmitted—through studies of birds in India. He followed the progress of the parasite's development, documented that it travels to the mosquito's salivary glands, and observed how healthy birds were then infected by a single bite. Manson was sick in bed when he received the news of these discoveries but nevertheless managed to rouse himself and travel to Scotland to announce them at a meeting of the British Medical Association.

In Italy, Grassi claimed that he had made discoveries similar to Ross's and declared that he had arrived at his first. Simultaneously, Bignami used mosquitoes purposely to infect human beings with their bites. The malaria circle—man to mosquito to man—had been closed.

From 1875, when Manson acquired his microscope, to 1898, when the Italians witnessed transmission to humans, Europeans and North Americans had conducted the first, widely publicized modern race to scientific discovery. Advances were reported on the front pages of newspapers worldwide, and most, if not all, of the leading researchers became obsessed with receiving credit for their work, not only in the world of science but in the public realm as well.

Many of the early malaria investigators worked with a fierce and sometimes nationalistic sense of competition, and they continue to do so. As they pressed their governments for funds, laboratory directors used patriotic appeals. And though they cooperated across national borders, sharing papers and even microscope slides, suspicion colored every exchange. It grew worse when dynamite king Alfred Nobel died in 1896 and news spread that his estate would begin awarding world prizes, with the largest monetary payments ever, for scientific discoveries.

The competition quickly reached its low point with the battle for credit between Ross and Grassi. Grassi was considered by many to be the preeminent medical zoologist of his time. He regarded Ross, who couldn't distinguish one mosquito from another, as an amateur and an interloper. When the Nobel Prize jury met to award its second round of medals, a joint honor was considered for both Grassi and Ross. But here again, jealousy arose. German scientist Robert Koch, who was himself locked in animus with Grassi, railed against the idea, and Grassi was dropped.

For the rest of their lives, Ross and Grassi would argue over primacy in several discoveries. Ross would be more widely recognized and honored. But he seemed never to feel secure of his place. At every opportunity, he attacked Grassi and inflated his own achievements.

Sadly, Ross's obsession with garnering every possible laurel eventually spoiled his relationship with his mentor. When Manson seemed to credit Grassi for his contributions, which were real, Ross turned cold toward his former teacher. By the end of his life, suffering from the effects of a stroke, Ross wrote an unpleasant book in which he harshly discounted Manson's science. Ignoring the evidence contained in the scores of affectionate letters that passed between them, Ross even suggested that Manson hindered rather than helped his malaria work.

In the decades that followed the unraveling of the malaria mystery, science refined all of the work done by Manson, Ross, and their generation. Dozens of different kinds of malaria parasites were identified and named. Many of these afflicted lizards and birds, but not people. Only four main species of malaria pathogens were found to attack humankind. The most common, but rarely fatal, is called *Plasmodium vivax*. Least common is *Plasmodium ovale*, which is restricted to West Africa and also produces a mild illness. A third kind, *Plasmodium malaria*, is found in isolated places scattered across the globe and, while it causes severe fever, it is usually not life threatening. The most vicious malaria, *Plasmodium falciparum*, can and does kill frequently, causing fevers and chills and destroying human lives due to anemia and capillary obstruction.

The evolution of drug-resistant malaria parasites followed the development of the medicines that eased the human victim's suffering. The first widely used treatment—quinine-based medicine—had been discovered centuries before anyone knew what caused malaria. The long history of the drug began in the early 1600s in Peru where, legend has it, Jesuit missionaries were offered powder from tree bark to cool fevers. By 1650, the powder was being used at Spiritu Santo Hospital in Rome. It reached England two years later and was a fixture in royal apothecaries throughout Europe by the 1680s.

In America, where malaria periodically roared through communities, especially along the Mississippi River, doctors experimented with various doses of the drug. In an 1850 study of malaria in the region, Daniel Drake, M.D., describes debates over the use of bleeding, purgatives, emetics, and a host of other compounds but unanimity over the benefits of quinine. "Everywhere, the sulfate of quinine is the popular remedy; and by all it has been found infallible."

Quinine's unpleasant side effects—including a sharp ringing in the ears—drove research into additional treatments. This process was accelerated during World War II when Japan occupied Indonesia and cut off the Allies' access to plantations where the bark was grown. A synthetic substitute—Atabrine—was developed in the United States, but an even better drug was made in Germany and, when it fell into American hands, it was copied by the army. This compound would eventually become chloroquine, an inexpensive, highly effective medicine that remains an important antimalarial drug worldwide.

The very popularity of chloroquine, which became a preventative for travelers as well as a treatment, inevitably led to widespread drug resistance. The first resistant parasites were recognized in South America in 1960, but soon they were seen around the world. Particularly problematic were regions where table salt had been chloroquinized in an attempt to eliminate infection by dosing entire populations. Over time, the parasites perfected their detoxification mechanism, and today the protozoa is something fully capable of killing its human host even when the best the pharmacologists have to offer is arrayed against it.

Since 1960, thousands of chemical compounds have been screened as malaria treatments, but none has been found to be as safe and powerful as chloroquine was in the 1950s. The newest drugs are expensive—as much as eight dollars a pill, compared with pennies for chloroquine—and present more toxic side effects. And in certain parts of the world the malaria parasite is resistant even to these. Millions still die of the disease every year.

The promise of better treatments rests, in part, on gaining a fuller picture of malaria's life inside the human body. What is known today shows that this pathogen has put millions of years to good use, evolving in a way that allows it to avoid destruction by the immune system.

The few dozen strandlike malaria sporozoites deposited by a probing mosquito reside briefly in the skin before they travel to the liver. Though little damage is done there, the liver is the place where the parasite transforms and initially multiplies. In a week or so, many tens of thousands more parasites will escape into the bloodstream before an immune response kicks in. With another trick of evolution, the parasite forms minute knobs on the blood cells, which help it avoid being cleansed by the spleen. It is then about two weeks after the mosquito's bite that the malaria victim begins to feel sick. But by this time, thirty or forty parasites deposited by a mosquito have become trillions inside the victim's body.

Uncovering the secrets of mosquito-borne malaria has also required us to learn more about the vector itself. It's not enough to understand that mosquitoes can deliver a pathogen to our bodies. To protect public health we must understand how this works. Is every mosquito dangerous? How many people will a single mosquito infect? How does an outbreak happen?

Ronald Ross laid the conceptual basis for answering these questions in a highly technical book that he wrote in the early 1900s. His ideas were later advanced by George Macdonald in the 1950s. Macdonald was director of the Ross Institute in London, and a man with varied interests. He spent many years in Assam, India, where he became a much-valued "tea taster," guiding the plantation owners in their efforts to blend their finished product. As much as tea drinkers benefited from his exquisite palate, medical science derived much more from his work on the central concept of contagion, now known as the "basic reproduction number" or BRN.

The BRN of a disease describes the number of additional infections that an originally infected person will generate under ideal conditions. A single person with measles, which is one of the most contagious infections, has a BRN of 12 to 14. The human immune deficiency virus, which causes AIDS, is far more difficult to transmit and has a BRN of just above one. A BRN

of one, of course is the minimum permitting the virus to perpetuate.

Add the help of mosquitoes, which are essentially flying
needles that extend a sick person's infectious range for miles, and
BRN soars. Malaria, for example, can have a BRN that exceeds
100—that's 100 cases traceable to a single carrier. In mosquito-
borne disease, three factors exert a powerful influence on the
equation that leads to the BRN. They are:

1. Vector abundance, or the population density of the mosquito that carries a disease.
2. Focused feeding, or the mosquitoes' tendency to bite people and nothing else.
3. Vector longevity. Will the mosquito live long enough to
   acquire a pathogen and then properly deliver it, via bite,
   to people?

Though anyone confronted with swarms of mosquitoes might
intuitively conclude that vector abundance is the most important
factor, it is actually the least powerful element in this equation. Focused feeding is a much more significant factor, and longevity is,
by far, the most important. An old mosquito is your worst enemy.

## YELLOW FEVER

With the puzzle of malaria solved, it was inevitable that the
mosquito's role in yellow fever, which stalked Africa and the
Americas and occasionally raged through European ports, would
be confirmed. Known as yellow jack among European sailors
and soldiers who visited the tropics, the disease causes fever,
jaundice, and hemorrhages that eventually kill a large proportion

of those infected. In its early stages, extreme muscle pain may affect the neck, back, and legs. Many people suffer neurological and psychiatric symptoms such as irritability and restlessness.

Yellow fever moves more quickly than malaria through a population, and once it attacks a person, it can bring death swiftly. Legends of ghost ships such as the *Flying Dutchman* were based on the real experiences of ship crews that were probably attacked by yellow fever during expeditions in Africa or South America. Amazingly, yellow fever has never struck Asia. Distance, quarantines, and good luck have probably kept the virus away. If yellow fever broke out in India, for example, where vector mosquitoes and their human hosts are exceptionally abundant, the loss of life would be cataclysmic.

The history of research into yellow fever turns on the work of Josiah Nott, of Alabama, who speculated early in the nineteenth century that mosquitoes transmitted both malaria and yellow fever. His conjecture inspired an experiment on human beings that would be impossible today, due to ethical concerns. In 1880 in Havana, Dr. Carlos Finlay allowed mosquitoes to ingest blood from yellow fever sufferers and then turned them loose on perfectly healthy human subjects. About one in five developed what Finlay considered to be a mild case of the disease. (In fact, he had not allowed enough time to elapse between bites for the virus, then yet to be discovered, to render these mosquitoes fully infectious.)

Finlay, whose earlier discovery that mosquitoes bite more than once was widely ignored, labored outside the scientific mainstream, and his scant observations were so easily criticized that he was dismissed as a crackpot. Historians would later suggest that Finlay was the victim of prejudice among scientists who could not accept that a significant advance could be made by a

Cuban. It didn't help his case that he wrote and spoke in a style that was considered florid and emotional by the British and German scientists who dominated biology at the time. Further damage was done to Finlay's cause by a number of scientific charlatans who also worked in Latin America. The most infamous of these was a doctor who had grown rich and been hailed as a hero in several countries in the 19th century before his so-called vaccine was proven to be ineffective.

Finlay's notion became more plausible after the mosquito's power as a vector of malaria and filariasis was finally confirmed by Ross and the Italians. The impetus to settle the question came from the Spanish-American War, which brought thousands of American troops to Cuba, where many of them succumbed to yellow fever. (Fewer than four hundred Americans died in battle while two thousand succumbed to tropical disease. Troops from Spain had suffered the same plight.) Once Cuba was occupied, alarmed American officials established a commission, to be headed by army doctor Walter Reed, which was charged with defending the U.S. Army from the fever's onslaught.

In the sketchy history that is most commonly repeated, Reed solved the yellow fever puzzle almost single-handedly. But few discoveries are made by a lone ingenious individual, and this was not one of the few. By the time Reed arrived in Havana, a young doctor named Jesse Lazear was already at work with two other members of the commission, Aristedes Agramonte and James Carroll, two physicians with broad experience in infectious disease.

Lazear was one of a few investigators who knew of Finlay's claim and had not rejected it. He had been eager to test it out. Reed was doubtful about this direction but supported Lazear's work even as the commission focused primarily on the idea that a bacillus, identified by Italian scientists, was the real culprit.

Almost all of Lazear's first attempts to use mosquitoes to transmit infection from patients to healthy subjects failed. The exception involved a soldier who had walked past Lazear's laboratory one morning and spied the young doctor on the porch holding two test tubes. One contained a mosquito. The other Lazear pointed at the sun, using the light to coax the insect into it. A conversation between the doctor and the soldier ended with the soldier declaring that he was hardly afraid of a tiny mosquito and volunteering to be bitten. He soon fell ill with a nonfatal case of yellow fever; but several attempts to replicate this successful transmission of the disease, and add to the proof, failed.

At the end of August 1900, Lazear was growing discouraged, and he shared his frustration, telling his colleague James Carroll that one of his infected insects was likely to starve to death because he had no willing subjects. Carroll rolled up his sleeves and volunteered.

The moment of Carroll's infection was reported in great detail in a paper written by Agramonte. In this telling, Carroll had actually taken the test tube that held the mosquito and pressed it against his skin. The sluggish insect eventually settled on his forearm and fed. Two days later, Carroll said he was not feeling very well. He blamed a swim he had taken in the ocean, where he presumed he caught a cold. Later that day, suspecting malaria, he drew his own blood and examined it for Laveran's organism.

On the following day, Lazear and Agramonte visited Carroll and both knew immediately that their friend suffered from something much worse than the common cold. Carroll was flushed, his eyes were bloodshot, and his lethargy shocked his colleagues.

Agramonte would later say that he and Lazear felt genuine panic about Carroll's condition. Lazear grew despondent over the fact that Carroll had been infected in his experiment. Together they nursed their colleague, sending daily cables on his condition to his family in the United States. Finally they were able to announce in a telegram, "Carroll out of danger." But with Reed back in Washington working on a report, and Carroll convalescing, Lazear and Agramonte were left to conduct the commission's work on their own. They were also charged with consulting in army clinics and local hospitals.

The conditions they labored under were far from ideal. Though the army had brought them to Cuba after yellow fever decimated the ranks, many officers were uncooperative. In one especially frustrating trip to a garrison where scores of soldiers were sick with the fever, Agramonte encountered a commandant and a camp doctor who refused to admit that an outbreak was under way. The garrison's doctor was treating the sick as if they had malaria, and Agramonte reported he was "under the influence of opium most of the time." The addicted doctor later committed suicide, and then Agramonte was eventually able to send medical teams to treat the sick.

One of the benefits that flowed from the outbreak was a steady supply of yellow fever patients, who aided the research done by the two doctors. Trying to disprove the idea that casual exposure to a patient made healthy people sick, they gathered soiled linen and clothes from yellow fever victims and piled them in a sealed room where volunteers slept for many nights. None of them contracted the illness, and neither did men who slept in the beds of soldiers who had died from the disease.

All of these findings led Lazear to believe he was on the verge of confirming Finlay's suggestion. But he must have feared being labeled a crackpot himself. In an excited letter to his wife he wrote, "Nothing must be said as yet, not even a hint. I have not mentioned it to a soul." Eager for further evidence, Lazear even considered letting an infected mosquito feed on his own blood.

On September 13, Lazear was in a yellow fever ward at a local hospital letting his mosquitoes feed on the blood of patients. As he held a test tube with a mosquito in it to one man's abdomen, a wild mosquito landed on Lazear's hand. Lazear thought about flicking it away. But he did not want to disturb the insect feeding in the test tube. He also hoped—not quite believed—that he had acquired some immunity to yellow fever through previous mosquito bites. So he let the wild one bite.

On September 15, Jesse Lazear told Agramonte that he felt a bit sick. Two days later he was bedridden, wracked by fever and colored by jaundice. For twelve days, he suffered with the sweats, fever, and delirium. Agramonte would later note that Lazear, during his more lucid moments, worried aloud about his family in the United States. Agramonte would write that the physical and emotional torture was only relieved "when the light of reason left his brain and shut out of his mind the torturing thought of the loving wife and daughter far away and of the unborn child who was to find itself fatherless on coming to the world." Jesse Lazear died of yellow fever on September 25, 1900. He was thirty-five years old.

Walter Reed was in Washington, D.C., when he learned of Carroll's illness and Lazear's death in a report he received from Carroll. His initial reaction to the scientific value of the experiments was negative. "I cannot say that any of your cases prove anything," he replied to Carroll. But Reed would quickly change his mind. On his return to Cuba, he conducted a thorough review of Lazear's work and became a believer.

Under Reed, experiments that conclusively implicated *Aedes aegypti* mosquitoes as the vehicle of transmission for the agent of yellow fever were conducted in insect-tight tents. Volunteers slept inside and provided blood meals for mosquitoes that previously had fed on yellow fever patients. For comparison, other volunteers slept in similar tents wrapped in blankets soiled with the "black vomit" produced by similar patients. Those who were exposed to biting mosquitoes sickened but not those sleeping in the soiled bedding.

On October 23, less than a month after Lazear died, Walter Reed announced as "our theory" that *Aedes aegypti* mosquitoes (then known as *Stegomyia fasciata*) were the vectors of yellow fever.

Another thirty years would pass before a yellow fever virus would be identified along with the fact that wild primates in the jungles of Africa and the Americas harbor a reservoir of the disease. But Reed's and Finlay's work allowed for the army to make war on Havana's mosquitoes and fully suppress yellow fever's spread.

Yellow fever virus still remains endemic in monkeys in the jungles of South America, where the primate population is large enough and so widespread that the virus can travel in wide circles, constantly finding new vulnerable victims. It will occasionally spread north through Central America and approach the southern border of the United States. These outbreaks are signaled by a progressively advancing wave of dying monkeys.

People are affected when a tree is felled, bringing the canopy down to them. An infected woodcutter may, thereby, return home with the virus incubating in his blood. *Aedes aegypti* mosquitoes would feed on him there, initiating a cycle of urban infection. After a person is infected, there is little even today that physicians can do. Vaccines are available and required for travelers to certain parts of the world. But local populations are stricken with the disease, and deaths do occur every year. Jungle yellow fever continues to threaten us with urban outbreaks.

## DENGUE

The pain that follows a dengue infection is so acute that the disease has long been known as "break bone" fever. The illness begins with a sudden high fever and rash, immediately accompanied by severe pain. The pathogen at work is one of four immunologically distinct kinds of dengue virus. Each confers immunity only to its particular type, which means a person can come down with dengue four times. The life-threatening version of this disease, known as dengue hemorrhagic fever, or DHF, results when a person encounters a second dengue strain. Severe internal bleeding is a part of this disease syndrome, and this is followed by shock. Death is common, particularly when the Type 2 virus strikes children who have previously been infected by another dengue type.

Dengue was the last of the major mosquito-borne diseases to be recognized. Prior to the nineteenth century, outbreaks in such far-flung places as Java, Egypt, and Philadelphia were mistaken for yellow fever and other illnesses. It was finally identified as a separate entity after a Caribbean epidemic in the late 1820s.

At the start of the twentieth century, with the mosquito firmly implicated in other tropical diseases, scientists began to consider mosquito transmission in the case of dengue. *Aedes aegypti* was conclusively identified as a vector in 1906. Other species would earn their place of blame later.

The complete picture of the dengue dynamic finds the mosquito squarely in the middle. The virus most likely originated in monkey populations, and monkeys remain the ultimate reservoir host in Malaysia. There and in the rest of the world, however, human beings maintain the virus, and mosquitoes ferry it from person to person.

Like yellow fever, dengue threatens to invade the United States from the south, through travel and trade. Unlike yellow fever, occasional cases are already occurring, mainly in Texas along the Rio Grande border with Mexico. However, the relatively high quality of housing there—most have screens, airconditioning—and the level of sanitation is such that transmission remains only transient. These outbreaks rapidly wane because the BRN in Texas is less than one.

The situation is very different in Puerto Rico, where conditions permit intense and perennial transmission of the dengue virus. The vector mosquito can be found virtually everywhere on the island, including in a flower vase that a conscientious maid had placed in my San Juan hotel room during a recent meeting of the American Society of Tropical Medicine and Hygiene. At least one of my colleagues at that meeting became painfully ill with dengue and suffered for weeks. Dengue's BRN in Puerto Rico appears to exceed two.

## ENCEPHALITIS VIRUSES

The mosquito's role as a bridge vector, connecting an animal reservoir host with people, is key to a number of viral illnesses that are marked by inflammation of the brain and the membrane that surrounds it. Lethal in some cases, these encephalitis viruses generally cause severe muscle aches, lethargy, and other symptoms in otherwise healthy children and adults. In the United States, the most prominent encephalitis viruses are carried by mosquitoes.

St. Louis encephalitis virus was first identified after an epidemic in and around the city of St. Louis in 1933. More than 1,000 people were stricken—additional cases were probably not reported—and 201 people died. More than 200 people were also diagnosed that same year in Louisville, Kentucky, Kansas City, and St. Joseph, Missouri. Among these, another 65 people died. When the epidemic began, the majority of public health officials suspected it was spread from person to person. Water, food, bad milk, even recently arrived zoo animals were blamed.

In the St. Louis case, Leslie Lumsden of the U.S. Public Health Service immediately suspected mosquitoes. Lumsden noticed that the illness occurred at a local hospital that was poorly screened, but not at an asylum that was well protected from mosquitoes. He also noted that cold weather brought an end to the outbreak. His mosquito deduction was roundly dismissed. As years passed, more epidemics and fatalities occurred in Ohio, California, Colorado, and Arizona. Finally, in 1942, the mosquito's role as a vector, carrying the virus to people, was confirmed. Research into the precise mechanisms of transmission continued for decades. Today SLE, as the virus is called, is found throughout the Midwest and has been seen as far east as upstate New York. No certain treatment is known for this disease; since its discovery, it has killed more than a thousand Americans.

Alongside SLE stand a number of similar mosquito-borne viruses that cause encephalitis. Western equine encephalitis can strike both horses and human beings. Veterinarians are often the first to notice that WEE is being transmitted in a locality, and they are responsible for alerting public health authorities. In the first half of the twentieth century, outbreaks of WEE, often involving hundreds of people, followed earthquakes in California. Most of the victims were infected when they moved out of damaged homes to sleep outdoors, where they were vulnerable to mosquitoes.

Eastern equine encephalitis (or EEE) is the most lethal of the mosquito-borne encephalitides. The infection first became evident as a human disease in the suburbs surrounding the city of Boston, Massachusetts, in 1938, in an outbreak that was coincident with the worst hurricane that ever struck the region.

The virus had been isolated from horses five years earlier. Little was known of its mode of transmission, however, and no one appreciated its potential as an agent of human disease. In the Massachusetts outbreak, a total of thirty-five people were hospitalized with this disease, all in a few weeks, and more than half died. The city of Boston was traumatized by this event, and fear of EEE continues to sear the public consciousness. EEE virus is transmitted in a broad arc along the eastern coast of the United States from Michigan, east through New York to southern New England, and south to Florida and the Gulf Coast.

The pattern of disease caused by EEE virus is particularly troublesome. Older victims of this infection generally die, while children who survive often become semicomatose, giving the disease its common name—"sleeping sickness." After recovering from the acute illness, many patients have long-lasting disability. Some experience "disinhibition syndrome," a condition that renders them likely to express any urge—including sexual urges—with action. Such people require constant supervision, and we calculated recently that survivors of this infection each cost society as much as $2.8 million in health care costs, long-term care, and lost productivity. The toll in terms of human suffering is, of course, impossible to calculate.

After EEE reappeared in Massachusetts during the mid-1950s, surveillance systems were established that were designed to detect incipient outbreaks in time for countermeasures to be organized. The most effective of these is a technique perfected in the early 1960s known as ULV insecticidal spraying, preferably dispensed from low-flying planes. ULV is the acronym for "ultra low volume," which was so named because the device can dispense concentrated insecticide. Previously, the insecticide had to be diluted as much as one part in ten. ULV changed everything because 90 percent of the cargo capacity of the plane would no longer be wasted on solvent, an enormous improvement. Now, insecticidal aerosols could be dispensed uniformly across broad regions. Although malathion remains the insecticide of choice, environmental and toxicological factors severely limit its use. A newer chemical, resmethrin, may be more acceptable.

The objective of ULV spraying is to generate an aerosol fog that remains airborne for about forty-five minutes and that will be lethal solely to insects smaller than a bee. Only a few ounces of insecticide are applied to each acre of land, and the particles are so small that they oxidize rapidly. The decision to spray generally depends on the discovery of a certain number of infected mosquitoes in a defined region.

EEE virus is maintained in many of the swamps that mark the countryside of the eastern United States. *Culiseta melanura* mosquitoes serve as the main vector. These "swamp mosquitoes" breed in underground bodies of water such as occur abundantly in swamps where the roots of certain trees are elevated above the water level in hummocks. A cavity forms there, just under the trunk of the tree.

*Culiseta melanura* has a curious history with man. Before asphalt tiles became generally available as roofing material, most houses were roofed with white cedar shakes, very durable boards that were split thinly from the most common tree growing in these swamps. As a result, white cedar was overharvested, and virtually all swamps in the northeastern United States were stripped of their trees. Swamp mosquitoes lost their habitat. They were seldom mentioned by the early entomologists and probably were exceedingly scarce.

With the advent of asphalt shingles, however, white cedar staged a major return during the twentieth century. And as a result, *Culiseta melanura* has become one of our most abundant mosquitoes. People who live near the swamps rarely see this mosquito because its bites are focused entirely on birds, the reservoir hosts of EEE virus.

The virus's journey from the bird reservoir to the human bloodstream can be taken via *Culiseta melanura*, but this would be extremely unusual. *Culiseta* circulates EEE among many kinds of birds but starlings are especially important. Starlings congregate by the tens of thousands in permanent communal roost sites. People acquire EEE mainly when they are bitten by another mosquito that serves as a "bridge vector" and bites both birds and people. *Aedes vexans* and *Coquillettidia perturbans* are two that will do the job.

Longevity makes *Culiseta melanura*, which transmits eastern equine encephalitis in the United States, a spectacular vector of human disease, even though it almost never bites people. This mosquito is the Methuselah of mosquitoes. I have kept them in cages in a laboratory for months while many other kinds of mosquitoes tend to die after a week or two. It feeds almost exclusively on bird blood. Indeed, it bites people so infrequently that scientists who have been bitten by this mosquito in the field have published reports noting the event in scientific journals.

Outbreaks of EEE occur after a wet winter and a long wet early summer followed by a heavy storm in late summer. These conditions permit the virus to amplify among birds before being carried to human or equine hosts. Regional public health authorities must monitor these conditions and follow the development of the epizootic (animal epidemic) before it bursts forth as an epidemic that destroys human lives. Any intervention requires an exceedingly difficult decision that takes into account environmental concerns as well as human safety.

Outside the United States, a wide variety of viral pathogens are transmitted by mosquitoes to people, causing encephalitic diseases of varying degrees. Japanese B is the most widespread of the mosquito-borne encephalitis viruses, occurring in much of Asia. Venezuelan equine encephalitis virus is similar to its North American cousins. West Nile virus may have originated in Africa and subsequently spread beyond that continent to Europe, part of Asia, and North America. All of these agents can kill, but in many cases healthy adults feel flulike symptoms and recover without treatment. Infections may even occur without symptoms.

## NOVEL DISEASE DYNAMICS

Though the era of momentous discoveries may be over— malaria, yellow fever, and other diseases have been explained— mosquitoes and the pathogens they carry continue to challenge and surprise us. Pathogens and mosquitoes are dynamic. They interact in the environment, change, and adapt. A fascinating example of these processes arose in the late 1970s when Rift Valley fever struck the Nile River Delta area. The Rift Valley virus is named for the enormous valley that begins along the border of Israel and Jordan and runs thousands of miles south through the Red Sea into Kenya and Tanzania. This was where the virus first was detected, in 1937, by scientists working in a famous laboratory located in Entebbe, Uganda.

Rift Valley virus can kill healthy, able-bodied adults by causing a terrible hemorrhagic illness that leaves its victims bleeding from every orifice. In this case in Egypt, hundreds of people were infected, scores died, and it was feared that the disease might eventually spread throughout the Mediterranean. No one, least of all officials in the Italian and Israeli governments, wanted to see a new vector-borne disease invade such a heavily populated region.

The outbreak was a mystery because it was then widely believed that the local house mosquito did not transmit the virus. And the best-known vector for the disease—the desert-dwelling *Aedes macintoshi* mosquito—was not present in the Nile Delta, where the fever had broken out. (*Macintoshi* larvae come to life after months or years of dormancy when rain floods a patch of the sands. Some hatch with the virus already in their salivary glands.)

Though the arrival of the virus in the 1970s was a tragedy for its Egyptian victims, it was also an opportunity. The investigation would significantly advance our understanding of the virus and its spread. And in a small way, it would contribute to the peace process in the Middle East. This possibility became apparent soon after I descended the stairs from an Ethiopian Airlines flight at Cairo International Airport.

I was met at the Cairo airport by Sherif El Said who, as usual, arrived in a chauffeured limousine. A well-connected gentleman, Sherif headed a mosquito research laboratory at Ain Shams University. Sherif was a scientist who also happened to have family in the Egyptian foreign service. Egypt and Israel had just begun normalizing their relationships, and Sherif's brother, an official in his country's Foreign Ministry, had told him that the time was right for a bit of mosquito diplomacy.

I was to serve as a go-between, helping to establish a three-nation project that looked at a variety of mosquito issues. But suddenly the Rift Valley outbreak had provided us with an urgent concern that demanded international cooperation.

Egypt had certainly not invited the Rift Valley virus and might not be able to prevent it from traveling outward in any direction, including eastward to Israel. Authorities in Israel had an obvious interest in aiding the Egyptians in understanding and containing the outbreak. They also had several scientists who had earned wide recognition for their work with mosquitoes, and they operated state-of-the-art laboratory facilities. A collaboration would be good for science, good for public health, and good for Egyptian-Israeli relations.

We were fortunate that Cairo was home to Harry Hoogstraal. Short, dumpy, with a cigar permanently fixed in the middle of his mouth, "Pasha Harry," as he was called, had spent decades in Cairo, through war and peace, as the main scientist of the U.S. Naval Medical Research Unit (NAMRU) that still is located there. The Cairo-based facility studied vector-borne diseases because troops in combat situations frequently encounter these pathogens.

(America's armed forces labs have long been interested in mosquitoes for reasons other than just defense. In the 1960s, the army's biological warfare experts at Fort Detrick, Maryland, had actually worked on the idea of using disease-laden mosquitoes as weapons against an enemy. Harry wasn't interested in this sort of thing. His speciality was ticks, but he was exceedingly knowledgeable about vector-borne disease in general.)

No one enjoyed food and drink more than Pasha Harry, and any engagement with him meant long nights at the dinner table in his home or the best restaurants in the city, complete with belly dancers and water pipes. But as much as he indulged at night, Harry was always hugely focused by day. He became an important colleague and an adviser in the work that led to the discovery of the virus in the local house mosquito. His cooperation, and the aid of military scientists in the United States, would prove vital to an understanding of the emergence of this pathogen in Egypt.

We knew from the beginning that the virus that was threatening the entire region had first appeared among farmers in the Aswan area, hundreds of miles south of the Nile Delta. It had probably come from nearby Sudan, where it is endemic among herdsmen who tend goats, camel, and cattle.

Mosquitoes could not have ferried the Rift Valley virus north, into the Nile Delta, because the distance was too great. It must have been introduced in the blood of a human being, or some other well-traveled animal, and then distributed locally by another mosquito. Members of Hoogstraal's navy group had found the virus in the common house mosquito, but there it reached a dead end. What was happening?

To help find the answer, we enlisted Mike Turell, the chief medical entomologist at Fort Detrick, where work could be conducted in a so-called hot suite where air locks, sterile gowns, and other precautions allowed for safer experimentation with dangerous agents.

Working with gerbils, because they served so nicely as experimental hosts for a variety of pathogens, we confirmed that blood-hungry *Culex pipiens* house mosquitoes became infected by the virus in the course of feeding, but they did not readily infect healthy animals during subsequent feedings.

In parallel experiments, we used needles to inject the virus directly into mosquitoes. With their very next meal, these injected mosquitoes were able to infect their mammal hosts. Somehow, the mosquito was much more capable of transmitting infections when it received the virus via needle than it was when it acquired it by sucking blood.

It was surprisingly easy to guess how this might happen and, more important, how Rift Valley virus may replicate in the wild in Egypt. When house mosquitoes feed on blood, they also ingest any pathogen that already is present in it. In the Nile Delta, one of the most common is the parasitic worm that causes elephantiasis. As you may recall, once sucked into the mosquito's gut these parasites—little worms called microfilariae—bore a hole, through which they escape into the insect's body cavity. They then migrate to the cells of its wing muscles, where they grow and mature. Ultimately, they escape by breaking out through the mouthparts of the feeding mosquito, swimming through the little puddle of mosquito blood that leaks onto the person's skin, and entering the human host through the hole in the skin created by the mosquito's feeding apparatus.

One piece of the puzzle of how the house mosquito could transmit Rift Valley virus was found. The mosquitoes that would most likely be responsible would be those that were infected simultaneously by the virus and microfilariae. The worms punctured the mosquitoes' guts—just as our needles had done in the laboratory. Thus wounded, a mosquito could no longer contain the virus within its digestive tract. Freed to travel, some virus particles ended up in the saliva, ready to be injected into the mosquito's next human victim.

In explaining how the common house mosquito may maintain the transmission of Rift Valley fever virus, we offered reassurance to Israel, Italy, and the various developed countries that ring the Mediterranean. Because they are essentially free of filariasis, they face little danger. Even if the occasional traveler or traveling sheep imported the virus, without the other infection, house mosquitoes would be quite impotent as a vector of this infection.

But the news for countries where filariasis is endemic was more sobering. A new dynamic in mosquito-borne diseases had been discovered, a dynamic that made the most common mosquito of all even more dangerous. In Egypt, the Rift Valley outbreak failed to become entrenched and ceased of its own accord. But it left behind a new fear.

# 6

# MAN AGAINST MOSQUITO

After Ross, with Manson's encouragement, demonstrated the mosquito connection to malaria, the world's sanitarians were poised to attack. At last they began to understand, scientifically, how their crusade to make the world clean and dry could make it safe from malaria. This knowledge could be used to persuade the public and government leaders to fund mosquito-reduction projects.

One of the world's most aggressive and successful anti-mosquito campaigns would be conducted in the United States by John B. Smith, head entomologist for the state of New Jersey. Smith began in 1900 with only a vague idea of the diversity of the species of mosquitoes that inhabited his state and with only a rudimentary understanding of how they lived, how they might be destroyed, and how the diseases that they carried might be suppressed.

Smith pioneered methods that would be copied around the globe. He identified the dominant species in his state, including several *Anopheles* mosquitoes that seemed likely to transmit malaria. He quickly discovered that these troublesome creatures breed in particular kinds of places; one of the most important breeding sites was the huge brackish marshland located north of the city of Newark, which is now called the Meadowlands.

What became known as "the New Jersey mosquito," *Ochlerotatus sollicitans*, developed in the brackish water that accumulated at the head of these marshes due to a heavy offshore wind or the spring tides that are driven by a full moon. *Ochlerotatus sollicitans* is a large, firm-bodied mosquito recognizable by the whitish stripe that runs down the length of the upper surface of her abdomen. She attacks from above, at any time of the day. Hundreds may descend on a person's head and upper body, driving their victim to retreat indoors.

Smith held a remarkably sophisticated view of the task he faced. He spoke of "mosquito control" rather than extermination, because he was sure that it was impossible to kill every mosquito in a given locale. However, this distinction was lost on the press in the region. Writers ridiculed Smith for even suggesting that mosquitoes could be dealt with, and the public at large generally agreed with the press.

Nevertheless, Smith received funding and planned a military-style campaign. His most powerful weapon was a motorized ditchdigger, which he used to slice drainage channels through thousands of acres of marsh. Mosquito "brigades," which Smith sent to every corner of the state, cut ditches and poured oil onto the waters where larvae developed. Barges equipped with dredging equipment pulled sand from river bottoms and pumped it to fill in swamps.

The effect on *Ochlerotatus sollicitans* was dramatic. Everywhere that Smith's crews worked, the mosquito's population crashed. Suddenly people could begin to enjoy the out-of-doors, especially the state's beaches. Cities such as Newark and Elizabeth saw development accelerate, and new neighborhoods were built in low-lying areas.

Local economies benefited enormously from the reduction in mosquitoes. In his report to state officials, Smith described a landowner who paid $50 to have a breeding area drained and then claimed that his acreage had increased in value by $10,000. Smith was thoroughly convinced that his efforts would make large areas suddenly more valuable. "Take the mosquito pest from Barnegat Bay and consider the resulting increase of visitors to that paradise for fishermen!" he wrote. "The increase in value in that territory alone would pay for all the work that would have to be done along the shore."

From the very start of Smith's campaign, the incidence of malaria in New Jersey declined. This was because his general attack on wetlands sharply reduced the breeding sites available to the eastern malaria vector, *Anopheles quadrimaculatus*.

By 1903, John B. Smith could note with some satisfaction that the "newspapers ceased to ridicule" and the citizens of New Jersey had come to believe that it was not only possible but absolutely necessary to reduce the density of mosquitoes. Throughout New Jersey, local communities began to form drainage committees. John B. Smith received visitors from around the world, all of whom were interested in his methods. And the successful control of mosquitoes in a village or town became a matter of civic pride. Communities competed to be the most mosquito-free.

At the time when Smith was cleaning up New Jersey's villages and towns, public health officials in New York City were developing and implementing a program that would work in densely populated areas. The city held a malaria control conference in 1901, which recommended attacks on breeding areas, screening for houses, and quinine treatment for patients. The plan was put into practice, with Staten Island getting special attention. By 1905, malaria was deemed to be under control in the city, and development on Staten Island began to boom.

Smith and his fellow antimosquito marauders in New York achieved results that are still paying benefits. Fly into Newark's airport, or across Staten Island, and you will see the ditches that continue to flush standing water and limit mosquito breeding.

To appreciate fully the value of the science that unlocked the mosquito's secrets, it helps to look at the spectacular failure of the French attempt to build a canal across Panama, which occurred just before the role of mosquitoes in human health was understood. Although history may hold stories of greater engineering defeats, it is difficult to think of a single one. During this project thousands of workers, including the entire crews of some ships, died of malaria or yellow fever. Investors—among them tens of thousands of middle-class people—lost the equivalent of $3 billion. And the reputation of a French hero was ruined. All of this trouble was caused, in large measure, by the mosquito.

A canal that would connect the Atlantic and Pacific was first imagined by the conquistadors who were the first Europeans to see the Pacific's eastern shore in 1513. But three hundred years would pass before even a railroad would be built across the isthmus. And no serious effort could be made on a canal until steam shovels and dynamite became the everyday tools of engineers.

The driving force behind the French canal project was Ferdinand de Lesseps, the William Gates or George Soros of his time, who had been the chief advocate for the construction of the Suez Canal. A man of legendary energy and forceful personality, de Lesseps was still fathering children and striding the world in search of conquests at age seventy-four when he offered to build a sea-level canal across Panama. After negotiating the rights to cut the canal, he crisscrossed the world to raise the capital. Construction began in 1881, with the canal's promoters, if not its engineers, declaring that the region's reputation for deadly disease was merely a fabrication of the project's critics.

The camps built to house the Panama Canal workers were probably the best in the tropics. But because mosquitoes were considered nothing more than a nuisance, screens were one luxury that was not indulged. In time, the barracks and even the hospitals built for construction workers became feeding areas for local mosquitoes. The French aided the insects' breeding as well, by decorating their compounds with lavish gardens where pottery rings filled with water were used to protect hundreds of trees from ants. These rings teemed with mosquito larvae.

The canal project suffered its first malaria deaths many months before a single ton of earth was moved. Surveyor Gaston Blanchett, who succumbed a month after completing a journey along the canal's path, was among thirty who died during the project's planning stage. Even if the project's directors had recognized the dire foreshadowing in these fatalities, there was nothing they could have done. Malaria was still a mystery associated with "emanations" from the earth.

Two years into the canal's construction, outsiders began to call attention to excessive deaths, but officials in charge did what they could to cover up the severity of disease among the workers. But by the end of 1884, a year that saw more than 1,200 men die, it was impossible to deny the toll. Accounts from the time suggest that in some parts of the jungle, two out of every three Europeans died from either yellow fever or malaria.

Like many others in their time, the canal project's managers connected sickness to some shortcoming in the moral life of each victim. Drinkers, gamblers, even executives who embezzled funds were all deemed to be particularly vulnerable to disease. One of the project engineers even came to believe that he could predict which of the newly arrived workers would sicken and die, based solely on the expressions on their faces as they came ashore. Another of the engineers said he would prove that only the immoral would die—by bringing his family to Panama. Not long after their arrival, his son, daughter, and son-in-law were all dead of yellow fever. He would lose his wife before succumbing himself.

Landslides, geology, civil unrest, and malicious rumors of de Lesseps's death eventually joined disease to defeat the French entirely. De Lesseps, who was eighty-four, reportedly fell into a melancholic kind of senility. Five more years would pass before he joined in death the estimated thirty thousand who had died before him in Panama. The financial scandal and legal proceedings around the failure of the French project would live well beyond de Lesseps. His son would go to jail, and the great French engineer Eiffel, who worked for the canal overseers, would pay an enormous fine for his role in the fiasco.

The half-built canal and the equipment the French left behind held real value only if someone came along to buy it all up and complete the task. After securing control over Cuba during the 1898 war with Spain, America possessed both a justification for a canal project and the means to do it more safely. The justification became clear to the American public when warships in the Pacific were unable to reach the Caribbean when they were needed. The means to complete it safely were perfected when the Americans drove yellow fever out of Cuba.

As was the case with Panama, America's relationship with Cuba was fraught with so much intrigue that even today the motivations and events that precipitated the Spanish-American War are in dispute. However, it is widely accepted that Cuba was viewed as a source of the countless yellow fever outbreaks in the United States, and one of the aims of the war was to impose some controls on this terrible export. Just prior to the war, experts gathered at a yellow fever conference in Montgomery, Alabama, to consider a resolution that urged the American government to take Cuba by force, with the sole purpose of attacking yellow fever. Feelings around this issue ran hot because, throughout the nineteenth century, no country paid a larger price for yellow fever than the United States, and it was widely accepted that the deadly disease was not homegrown but rather delivered from the Caribbean, especially Cuba.

War's end found Cuba free from Spain but not yet independent. The U.S. Army was in residence as an occupying force, and though the Spanish Army and Navy were gone, U.S. troops were nevertheless threatened. Roughly 80 percent of the Americans sent to the island came down with yellow fever, and Congress dispatched the commission, headed by Walter Reed, that eventually confirmed the mosquito connection.

As Reed, Lazear, and the others conducted their experiments, an army major named William Crawford Gorgas had been struggling in vain to use ordinary sanitation methods against the epidemic. A straight-backed man with prematurely white hair and a thick mustache, Gorgas was an imposing figure who was comfortable with power. The son of an army officer (U.S. and later Confederate), Gorgas was in Charleston when the first shots of the Civil War were fired at Fort Sumter, and he viewed Stonewall Jackson's body when it lay in state in Richmond. After the war he attended university in New Orleans, where he got his first look at yellow fever. He studied medicine and became an army doctor.

In Cuba, Gorgas wielded unchallenged military authority, and he used it to turn Havana into what may have been the cleanest major city on earth. But rather than subside, the yellow fever outbreak worsened as more and more soldiers arrived to become fuel for the virus.

Once the Reed group had established that a local *Aedes* mosquito carried the yellow fever pathogen, Gorgas considered various methods for preventing the disease. One involved "vaccinating" men with a bite from a recently infected mosquito, inducing a mild case that would produce immunity. Unfortunately, gathering virus-bearing mosquitoes was difficult, and they had a tendency to die. (One biting mosquito that Gorgas affectionately named "Her Ladyship" was so precious that her death was followed by a mock funeral.) Over time, this inoculation scheme was found to be unreliable, and Gorgas had to find another way to prevent yellow fever.

Because sick men provided the reservoir in which mosquitoes could acquire the virus, Gorgas came to see them as a point where the chain of infection could be broken. He ordered that patients be isolated in fully screened buildings for the duration of their illness. At the same time, Gorgas went after the mosquitoes. Brigades of soldiers patrolled the city and its suburbs in search of containers of standing water where *Aedes aegypti* might breed. They emptied or smashed every container they found.

Ponds were oiled every week. If larvae were found on any property, its inhabitants were punished. In five months, yellow fever was gone.

The elimination of yellow fever from Havana made Gorgas a hero. In 1904, when America took up the task of completing the Panama Canal, scientific experts urged that Gorgas be appointed to the commission overseeing the project. However, his reputation for painstaking thoroughness made many fear that he would slow down the canal project. He was denied a place on the commission, and even when the army appointed him to serve in the canal zone, his work was regarded as more of an annoyance than a benefit. Then came the rainy season, and the mosquitoes.

The first cases of yellow fever among the Americans in Panama were diagnosed in November 1904. The first person to die was the young wife of one of the administration's office staff. Within three months, scores had died. Funeral processions followed the railroad line, which was also the route of the canal. Fear of this dread disease grew among engineers and laborers alike. Ships that arrived carrying new workers departed with even more men fleeing yellow fever.

Working against both the mosquitoes and the many officers who disliked him, Gorgas instituted the same kind of anti-vector measures that had worked in Havana. He cleared the French gardens and their water receptacles away from living quarters and hospitals and installed screens wherever the sick lay. He spread larva-killing oil on larger breeding sites and sent squads out to destroy containers that might hold rainwater. He even sent soldiers into homes and barracks to fumigate rooms and to swat adult mosquitoes, one by one.

Gorgas made slow progress, and officers and men came to call him a crank. His style left much to be desired. Panama was not Cuba. A great many civilians were engaged in the canal project, and they did not appreciate Gorgas's tendency to order them about. But Gorgas also ran up against a commanding officer in Admiral John Walker who, even after Havana, considered the mosquito-disease connection so much "balderdash." Walker was so terrified of charges of graft and corruption that he resisted any spending that didn't seem directly related to moving earth. Gorgas was forced to devote much of his time to pleading for funds, with little result, save for a sour relationship with Walker.

Eventually Walker and the commission decided that Gorgas needed to be replaced. (One member of the commission went so far as to tell him, "Everyone knows yellow fever is caused by filth.") But the hero of Cuba was not without allies. President Theodore Roosevelt himself decreed that Gorgas would stay. And when a new chief engineer was appointed, the mosquito-killing major gained all the power he would need.

By 1906, Gorgas managed to eliminate yellow fever from the canal zone. Malaria, which relied on a mosquito vector that was much harder to suppress, lingered, but at a much lower level. During the decade it took for America to finish the canal, roughly 2 percent of the workforce was hospitalized at any given time. This compared with 30 percent for the earlier French project.

Despite Gorgas's success in the canal zone, his critics cited the cost of mosquito control—two dollars per man, per year—suggesting it was out of proportion to its reward. Gorgas countered that he saved far more in hospital costs than he spent on mosquito control. And he pointed out that the American public would likely have withdrawn support for the canal if the death rate had been similar to that suffered in the first attempt to cut the isthmus. Eventually Gorgas would be fully recognized for his achievement, promoted to the rank of major general, and be named surgeon general of the army. King Edward VII of England, where tropical medicine remained a preoccupation, knighted him in recognition of his achievement.

Although the American experience in Panama may have suggested that an extensive territory could be made safe from mosquitoes and their associated diseases, this goal had not actually been achieved. The canal zone was a narrow strip of land barely forty-five miles long and fifteen miles across, and most of its acreage was uninhabited. The communities in the zone had been rendered free of mosquito-borne infection, but malaria and anopheline mosquitoes were still found nearby, along with the vector of yellow fever virus.

The basic lesson of Panama was that a limited area could be rendered free of mosquito-borne infection for the benefit of a community of foreign invaders. This reinforced the tendency of Europeans to rely on segregation to protect their troops, administrators, and settlers in other tropical regions. At the turn of the twentieth century, Alphonse Laveran helped write a guide for French colonists that advocated many of Gorgas's methods, especially screening and drainage, but he also advocated separate communities for natives and Europeans. The British schools of tropical medicine supported this approach with erroneous reports declaring that only "natives"—and not European settlers—carried malaria parasites in their blood. As a result of this misconception, local governors enforced various rules against Europeans mingling with, or even settling near, local people.

In many parts of Africa and India these policies made life safer and healthier for outsiders. Mine operators, plantation owners, and others pointed to reduced rates of disease among Europeans. A bridge company hired to span the Zambezi River in the early 1930s reported proudly that mosquito control had eliminated disease among workers brought from England. Under these conditions, the bridge was actually finished ahead of schedule.

On a philosophical level, Europeans and Americans generally saw their attacks on mosquitoes and the pathogens they transmitted as part of the "white man's burden" of bringing their kind of civilization to the tropical world. Ronald Ross expressed this chauvanistic notion when he wrote that malaria "strikes down not only the indigenous barbaric population but, with greater certainty, the pioneers of civilization—the planter, the trader, the missionary and the soldier. It is therefore the principal and gigantic ally of Barbarism."

The efforts of Ross and his peers to protect Europeans from mosquitoes made life for locals living nearby much more challenging. Those who lived nearest to the Europeans may have experienced a reduction in disease, for example, but this could disrupt their immunity and leave them vulnerable to future outbreaks. This was especially true in communities just within the line where mosquitoes were kept in check. People who lost their immunity would contract malaria if they happened to travel to see relatives or for work.

This is precisely what occurred in the mines of West Africa, which yielded tons of gold for Great Britain. Hundreds of black miners died of yellow fever during outbreaks that were mostly ignored by British authorities. In other areas, colonial policies forced people to relocate—sometimes en masse—with the result that people were exposed to new mosquito-borne pathogens. Finally there was the problem of men who labored in one region

and carried viruses or parasites home to another, thereby sparking outbreaks that started with their own wives and children.
This happened many times with malaria as foreign variants were
introduced into isolated communities with devastating results.

Other changes wrought in the African environment by Europeans contributed to disease among local people. As trees were
felled, yellow fever virus was brought to ground level, where
woodcutters could be infected. They brought the virus back to
their home communities. At the same time, millions of man-
made items discarded by Europeans became breeding sites for
mosquitoes. The vector also bred in the ditches that ran alongside roads that the newcomers pushed through the wilderness.


As Europe and the United States fought mosquitoes abroad, they
also carried out vigorous domestic campaigns, targeting malaria
especially. The disease was common in Scandinavia, the British
Isles, continental Europe, and in the American South and Midwest. In these lands, literally thousands of drainage projects were
established to reduce the breeding areas for mosquitoes. Public
works were even more effective when combined with the universal use of screening in homes and public buildings. Finally,
the widespread use of quinine destroyed the human reservoir for
the malaria parasite.

By the year 1914, cities as distant as Rome and Rio de Janiero
could claim to have reduced malaria mortality to a handful per
100,000 residents. In North America, malaria had been effectively pushed out of Canada and the United States, as far south
as Washington, D.C. Although most of the American Midwest
was clear, the deep South remained affected. The warm climate
and wet landscape contributed to this problem, but it was mainly

poverty that kept the region malarious. The worst recorded rates of malaria infection were in Mississippi, where annual deaths from malaria were roughly 1,000 per 100,000 persons. In Sunflower County, Mississippi, possibly the worst case in all the United States, the rate was three times as high.

The effects of this illness on local economies was well understood. In North Carolina, textile mill operators estimated that worker efficiency was cut in half during the four months of the year when malaria was most pronounced. Southern railroads reported that one-third of the sick days claimed by their workers were due to malaria. In a survey of seventy-four Louisiana farm families, the U.S. Department of Agriculture documented 1,066 days of lost labor in the span of four months due to malaria.

During World War I, antimalaria campaigns were finally focused on the South, where poorly funded local governments had been reluctant to pay for antimosquito programs. The U.S. Army played a major role in this effort by attacking mosquitoes in the areas that surrounded every garrison, arsenal, camp, and base on American soil.

After the war, the Rockefeller Foundation, which Gorgas had joined upon retiring from the army, funded experimental projects that showed how malaria could be suppressed at a much lower cost—eighty-one cents per person, per year—than was previously assumed. Local governments began paying for malaria programs, and the incidence of the disease began a steady decline. This was accomplished with a variety of now tried-and-true methods including drainage projects, screening, the oiling of breeding sites, and the wholesale distribution of quinine to kill parasites in the human reservoir. Little Sunflower, Mississippi, saw quinine defeat its malaria problem in a year's time.

Remarkably, even the Great Depression could not halt the

antimalaria effort. Roosevelt's Works Progress Administration dug thousands of miles of ditches and drained hundreds of thousands of acres in the South. The drainage work was carefully engineered to disrupt the breeding of the presumed vector, *Anopheles quadrimaculatus*.

*Anopheles quadrimaculatus* breeds at the margins of large bodies of water, where water, air, and solid plant material come together. The more complex the surface, the larger the opportunity for breeding. The Tennessee Valley Authority in the Southeast devised an extraordinarily effective strategy for disrupting conditions in these breeding areas. The banks of the reservoirs were cleaned of vegetation and dressed at an angle precisely designed to strand the larvae when engineers drew the water level down by regulating the flow at dams. Other wet areas were drained or filled, and a rigorous procedure for malaria surveillance and response was put in place. Any case of malaria that was detected was immediately investigated and remedial measures instituted.

Besides managing the flow of water, the TVA prescribed and even installed screens on the windows and doors of homes throughout the region. (Today, many states continue to legislate their use, and the banging of the screen door has become the characteristic sound of summer in the countryside.)

The American experience showed how the combination of finely targeted attack on mosquitoes, simple interventions like the installation of screens, and the natural human patterns of migration was effective against malaria. Once the impact of the vector was sufficiently reduced, and the sick received treatment, the parasites resided in fewer and fewer human bodies. Eventually it would be the parasite, not the mosquito, that would disappear. Today the United States is home to plenty of *Anoph-*

*eles* mosquitoes, which are quite capable malaria vectors. But malaria itself is almost never seen. When minor outbreaks of a dozen or fewer cases occur, they are invariably traced to a visitor from abroad who carried the parasite.

As nations mounted offensives against mosquitoes and diseases they carried, officials also began to see the need for defenses against the pathogens that lurked on foreign shores. Faster ships, new routes made possible by the Panama Canal, and an increase in air travel heightened the fear of introduced disease. Many countries harbor mosquitoes that can quickly spread a virus or parasite once it is introduced. Again, the most nightmarish scenario anyone concerned about mosquitoes could imagine involved the introduction of the yellow fever virus into Asia, where hundreds of millions of people are vulnerable and billions of mosquitoes available to transmit the infection.

Of course, many countries also live in fear of the second half of the disease dynamic: the mosquito vector. Major efforts were made to police borders against mosquitoes. In many locales, trained experts inspect breeding areas near points of entry—harbors, railyards, airports—on set schedules. But in March 1930, Raymond C. Shannon was simply taking a Sunday morning walk near his home in Natal, Brazil, when he looked down and happened to spy a strange larva wriggling in a puddle. An entomologist on loan from the Rockefeller Foundation, which had assumed a worldwide role in fighting mosquito-borne illness, Shannon soon confirmed that he had found trouble in that puddle. He had found *Anopheles gambiae*.

Originating in Africa, the *Anopheles gambiae* complex of mosquitoes includes the most efficient malaria vectors on Earth.

They adapt very well to life in human communities and feed almost exclusively on human blood. These mosquitoes are well established in the West African regions that are due east from Natal, across the South Atlantic. Because flights between the two regions were then rare, it was likely that they had come over on one of the fast French destroyers—called *Avisos*—that were used for mail delivery at the time. The insects probably traveled in adult form and simply flew ashore.

Once on Brazilian territory, *Anopheles gambiae* had the good fortune to find hay fields that had recently been created with the installation of a system of dikes. If not for the dikes, the mosquitoes would have found only salt marsh, which is not suitable for their breeding. Instead they encountered perfect conditions and established a colony. With its beachhead in Brazil, *Anopheles gambiae* was poised to spread throughout South America and even push northward.

Though the larger, hemisphere-wide threat was recognized from the start of the *Anopheles gambiae* invasion, the original infestation became the concern of the small city of Natal and the state of Rio Grande del Norte. To the dismay and frustration of Raymond Shannon and his Rockefeller colleagues, local officials seemed barely interested. Shannon and others recommended that the dikes be opened to allow salt water into the breeding area. Some hay might be ruined, but they figured it was a small price to pay. One official did not agree, however, and the dikes remained closed.

A small malaria epidemic arose as the experts and public officials argued over the hay fields. In the small community nearby, more than a hundred cases were noted in the month of April. Enough people died that the government had to supply food for some families who had lost their wage earners. In January 1931, the outbreak exploded into a fierce epidemic with ten thousand people falling ill. An aggressive campaign was mounted to fumigate houses, lay oil on some breeding areas, and drain others. Before the year's end, malaria was in retreat from the city.

Malaria and its vector did leave the city of Natal, but they were not driven out of northeast Brazil. In the years that followed, they would appear in small villages, some as far as 150 miles from Natal, do a little killing, and then be repelled by a local control effort. Forty people died in the fishing village of São Bento. More than a hundred were killed in Taipu. When *Anopheles gambiae* hit São Gonçalo, all but a handful of the town's residents fled.

In retrospect, scientists would say that Brazil's authorities should have recognized a larger problem in the pattern of small outbreaks. But the national health service was preoccupied with famine and disease caused by an extended drought. Refugees from stricken areas were settling in camps, all of which required water supplies and sewage disposal to prevent outbreaks of infectious disease. These challenges alone were more than the government could handle. A possible problem with an African mosquito that troubled just a few isolated villages didn't seem like a priority.

Except for causing those minor outbreaks, *Anopheles gambiae* allotted Brazil five years of peace and then struck hard. In 1938, the largest malaria epidemic in the history of the Western Hemisphere struck 100,000 people, killing between 14,000 and 20,000. (Because many rural deaths were not recorded, no firm figure was ever determined.) Affecting towns and villages across a two-hundred-mile-wide swath of the countryside, the outbreak made entire villages sick. A survey in the town of Baixa Verde indicated that half the houses were empty due to panic.

Among the 1,060 people remaining in town, 1,012 were sick. In Santa Luzia, all 400 residents were ill.

The epidemic's effect on business, agriculture, and even state services was profound. In some areas, 70 percent of the cotton crop was never picked because the harvesters were too sick. Hospitals ran out of beds. Pharmacies exhausted their supplies of drugs. Power companies lacked the necessary workers to maintain electric service. When the food delivery system broke down, many people died for lack of nourishment even though their infections were mild.

Brazilian doctors were shocked by the virulence of the disease they encountered and the rapid pace of its spread. They were accustomed to the malaria that was endemic to Brazil and was spread rather haphazardly by local vectors. People probably were multiply infected by these efficient vector mosquitoes, which made the resulting disease more severe.

From a public health perspective, the speed and extent of a malaria outbreak are astounding. In many places where epidemics have occurred, more than half the residents are stricken with an illness that makes it impossible for them to function. The result can be especially devastating in agricultural economies where planting or harvesting depend on manual labor and must be completed during a particular season. In these cases, the disruption in the pattern of everyday life becomes more devastating than the disease.

Taking into account the seriousness of the situation, Brazil's president, Getúlio Vargas, created an emergency antimalaria service, staffed it with four thousand workers, and placed it under the command of Fred Soper, a Rockefeller associate. It is hard to imagine a man whose personality was better matched to the task. Meticulous, driven, perhaps even fanatical, Soper was the third of eight children raised by a pharmacist and his wife in a small Kansas town. His most vivid memories of childhood included terrible bouts of illness, which likely influenced his choice of career. Soper was five foot eight, but he had the presence of a giant. Uniformed in a business suit, his thin mustache trimmed to a razor's edge, Soper spoke with an unwavering certainty that demanded obedience from underlings. A 1969 book, *The Plague Killers*, noted that Soper "seemed equally capable of browbeating man or mosquito."

Working with the authority of a martial law commander, Soper quickly established a headquarters in Fortaleza with laboratories, a cartography department, and training facilities. He then divided the epidemic area into small zones and assigned an antimalaria brigade to each zone. Empowered to inspect every property and enter every building—by force if necessary—the brigades attacked with larvicides and indoor insecticides. Their favorite weapon was a spray called Paris Green (copper acetoarsenite), which killed larvae in water. Some drainage projects were established to eliminate standing water, and local citizens were taught to cover the shallow wells they dug to obtain water.

In addition to his war on *Anopheles gambiae* in the region where malaria had broken out, Soper conducted reconnaissance in a wide area, looking for both the disease and the vector. When twelve *Anopheles gambiae* were found in a car traveling out of the infested area, Soper recognized that highways could aid the mosquito's spread the way rivers had for centuries. He established more than thirty "de-insectization posts" on the roads. When drivers reached the barricades, they were required to submit their cars and trucks for fumigation. Soper did the same to trains, at seven different checkpoints. Ships were boarded, inspected, and fumigated. Airplanes were also treated—mosquitoes were found on Pan American Airways planes arriving in Brazil from Africa.

At its height, the African mosquito's invasion force held well over eighteen thousand square miles of territory in three Brazilian states. In a single year, Soper managed to drive it out of all but two small inland towns. These infestations would soon be taken care of by local authorities, and complete victory over *Anopheles gambiae* was announced. Soper and his four thousand brigadiers had, it seemed, saved the Americas from the world's most dangerous malaria vector.

Soper was rewarded with medals, citations, and the gratitude of the Brazilian people. Eventually a gold medal was struck, with his profile on one side, to honor his leadership of the Pan American Sanitary Bureau, which would fight disease throughout the hemisphere.

In Soper's time, many vector-borne disease experts believed that the Brazilian experience had answered some long-standing questions about mosquitoes. It confirmed that species can adapt rather quickly to new environments and then thrive. This knowledge gave the public health experts of the world pause. But Soper's army also discovered that with extraordinary effort, a fairly large region could be rid of a mosquito pest. For the first time, many scientists were coming to believe that complete eradication, not just mere control, was an attainable goal.

Soper's victory came to be regarded as one of the greatest public health achievements in history. In his later reports on the project, he detailed his campaign in cool language, emphasizing that spraying was the key to victory. In his conclusion, he argued forcefully for immediate action when a mosquito crisis arose, and an assault on the vector rather than cautious control of the disease. Soper, the world's great insect killer, had come to see the mosquito as a formidable but vulnerable enemy that should be subjected to an all-out war. He believed eradication was possible and preferable. Soon he would have the ideal weapon for the battle.

PART THREE

# THE BALANCE

# 7

# THE GREAT MOSQUITO CRUSADE

In another time, Fred Soper might have enjoyed even more of the hero's rewards. After all, he had exceeded Gorgas's accomplishments in both Panama and Cuba, demonstrating that an alien mosquito could be driven from the shores of an entire continent. But this achievement came just as the world was plummeting into global war. Soper and most of the other mosquito men would leave science for the moment in order to serve their respective countries. Many would go directly to the fields of battle to fight infectious disease.

During World War II, educated commanders understood that illnesses, especially those carried by mosquitoes, could have a crucial effect on the outcome of engagements. History offered plenty of examples. Long before anyone knew how malaria is transmitted, Napoleon had purposely relied on its presence on the Dutch island of Walcheren to defeat a British force of twenty-five thousand. Malaria was a factor in numerous World War I battles in the Balkans and Africa. One infamous illustration of the effects of this disease came when a malaria-weakened

British Army in Macedonia tried to summon nearby French units for relief. The French commander's response—"Regret that my army is in hospital with malaria"—immediately found a place in the annals of war. The British might have been over-run for lack of reinforcements but for the grace that malaria is not partisan. It turned out that the German enemies were sick too and unable to strike.

Mindful of history's lesson and determined to send the health-iest possible forces into battle, the Americans in World War II established a malaria control office under the surgeon general of the army and staffed it with many of the world's leading experts on vector-borne diseases. Decades later, Paul Russell, M.D., would recall how he was initially assigned to the malaria office in Washington. Soon, however, he was summoned to the South Pacific by Douglas MacArthur, whose forces were pinned down, not by enemy fire, but by mosquitoes. Russell, who previously traveled the world fighting malaria for the Rockefeller Foun-dation, had been made an army officer and was subject to Mac-Arthur's orders.

"When I came in [to MacArthur's office] he got up and said, 'Doctor, I have a real problem with malaria,' " recalled Russell.

MacArthur told Russell that at any given time, one-third of his fighting men were in the throes of malaria, one-third were recovering, and only the final third were truly fit for combat. If this problem were to persist, said the general, the war would be a very long conflict.

On MacArthur's orders, army staff guided Russell on a fast tour of the Pacific theater. He discovered that men in the field were not taking prescribed doses of the malaria drug Atabrine, which served effectively as a prophylactic, and that antimalaria squads were not in place. Apparently mosquito fighters had been trained and their supplies were ready. But they were considered so peripheral to the war effort that they couldn't get transport to the front.

Russell made just two recommendations. First, commanders had to be made responsible for their men taking Atabrine. Second, antimalaria squads and their equipment must be given first priority for transportation. MacArthur immediately issued both of these recommendations as direct orders. Within weeks, the antimosquito squads were at work throughout the Pacific theater and most troops were complying with the order to take their medicine. Malaria rates among U.S. troops declined. Soon, the American forces were healthier than their foes, who never were as disciplined about malaria prevention.

In the Pacific, the military's antimalaria effort became a part of the war culture. Posters and cartoons in the Army newspaper *Stars and Stripes* constantly reminded soldiers to keep their sleeves rolled down and Atabrine at hand. (Malaria Moe and Anopheles Ann were featured characters.) Armed Forces Radio broadcast so many antimalaria messages that troops began calling it the Mosquito Network.

As Russell fought malaria in the Pacific, Fred Soper confronted epidemic typhus in Africa and Europe. Spread by body lice, typhus is known to erupt and then rage through refugee camps and other places where war disrupts sanitary conditions and forces people to wear the same clothes for weeks on end. Because the body lice that transmit this infection lay their eggs in a person's clothing, these infestations multiply most intensively among refugees, soldiers engaged in trench warfare, and others living in close quarters. Millions of soldiers and civilians had suffered from typhus during and after World War I. Roughly one in four had died.

Perhaps the world's all-time champion at killing tiny animal pests, Soper joined a team in Africa that was under orders to find a better way to kill the body lice that served as vectors for the agents of this disease as well as relapsing louse-borne fever. (Unlike Russell, Soper declined a military commission.) Much of the typhus work in Africa was done at a squalid Algerian prison where the malnourished inmates were vulnerable to lice and the disease that they transmit. Soper would later recall how he had to steel himself to work among the more than one thousand prisoners, many of whom seemed to be starving to death before his eyes. "They were poorly fed and kept at hard physical labor," he would recall. These men served as the human breeding ground for body lice. Soper's group searched the laundry for the creatures and thus played an important role in a historic discovery.

In the summer of 1943, one of Soper's colleagues returned to Algeria from Orlando, Florida, where he had acquired five pounds of a new chemical—then called Neocide—that was being tested at a federal laboratory. In Algeria, Neocide was mixed with an inert powder from a local cement factory and sprinkled onto clothes infested with lice. Like many compounds, it killed the lice. But unlike the other insecticides, which had to be applied weekly, it kept on killing for months. And it seemed to cause no harm to other living things. It was the breakthrough for which Soper and company had been hunting.

Ironically, the compound that had killed the prison lice had existed for decades. Swiss scientist Paul Muller had crafted Neocide, to be used as a moth killer, out of a chemical that had been invented in Germany almost fifty years before. For the Allies, however, it was a new and potent weapon. And soon after its effectiveness was demonstrated, British and American factories began turning it out by the ton under a new name—dichloro-diphenyl-trichloroethane. Everyone referred to it by the initials DDT.

Axis prisoners of war in a camp in Casablanca were among the first to be deloused with DDT, but the new chemical proved its real worth in Naples, where the fleeing Germans had destroyed sewer and water systems, and the Allies encountered refugees who were teeming with lice and infected with typhus. With Soper taking charge, the army dusted fifty thousand people per day with DDT. In a matter of weeks, the outbreak was over.

As Soper and his delousing squads cleared typhus out of Naples, the U.S. Army proceeded to battle through the rest of Italy. Fleeing German troops destroyed the tide gates that had been installed in the Roman *campagna* to prevent accumulations of brackish water, the favored breeding ground for *Anopheles labranchiae*, the local carrier of malaria. Water rose in thousands of acres of marsh, which greatly expanded the breeding opportunity for this mosquito. Allied soldiers were then plagued by the insects that emerged, and became infected with malaria. About eight thousand soldiers fell ill just prior to the start of the five-month battle of Monte Cassino.

Over the course of the war, many more soldiers were evacuated for malaria than for the treatment of wounds. In the Pacific, the outcome of battle was often decided by the malaria count on one side or the other. The Allies were generally better than the Axis powers at policing malaria. But there were times when the sides were so near each other on the battlefield that mosquitoes delivered parasites to GIs that had developed from other parasites that the mosquitoes had imbibed from the bloodstreams of nearby Japanese troops.

In the waning days of the war, army officials began to believe that DDT could become the ultimate weapon for the mass destruction of the mosquito. It was used on Okinawa, where a couple of battered DC-3 airplanes were equipped to dust much of the countryside. The war ended before this chemical could get a serious combat trial. But when peace arrived, civilian authorities also recognized that DDT might be the answer to a host of insect-related diseases. The big new chemical gun offered the greatest hope to those who confronted mosquitoes and had dreamed of following Soper's eradication scheme worldwide.

Years later, Soper would describe the triumph of his eradication model with as much delight as his flat, Kansas accent would allow. For decades, the very thought of species removal was derided by scientists who were certain that nature wouldn't allow it, no matter how hard man tried. Many malariologists doubted that eradication was possible. But with Soper's success, and DDT's power, those who advocated killing every last suspect mosquito seemed to have won the argument.

The optimism of the exterminators could also be heard in the lectures that Paul Russell, by then a visiting lecturer at the London School of Tropical Medicine, would give in the early 1950s. "This is the DDT era of malariology," declared Russell. "For the first time it is economically feasible for nations, however underdeveloped and whatever the climate, to banish malaria completely from their borders."

Such was Russell's public position in 1953, despite the fact that DDT had already begun to show some weakness. In fact, the majority of mosquito fighters would remain loyal to the wonder pesticide long after it fell short of its promise.


When Fred Soper proved in Brazil that a mosquito could be eliminated from a vast region, he also set a standard for success that would best be described as "zero tolerance." If even a few mosquitoes are left alive at the end of an eradication campaign, Soper argued, the effort is a failure. This is especially true with anopheline mosquitoes and malaria. Only complete annihilation of the vector would work, he said. The concept of "species sanitation" was born.

One of the first large civilian DDT programs proved Soper's point. The target was the Italian island of Sardinia, with a population of about 1.2 million. Long endemic, malaria had flared into a raging epidemic there during the war. In 1946, more than ten thousand new malaria cases were recorded on the island.

To confront malaria in Sardinia the Italian government and the Rockefeller Foundation launched an enormous campaign, which at its peak involved twenty-five thousand field workers, five fixed-wing aircraft, two helicopters, hundreds of jeeps and trucks, and a network of field offices. Their enemy was a mosquito named *Anopheles labranchiae*, a soft-bodied, brown-colored mosquito with four dark spots on each of its wings. The "trident" appearance of its head, due to a pair of prominent palps—sensory appendages—that are as long as its proboscis, marks this mosquito as an anopheline.

The army assembled for the job attacked *Anopheles labranchiae* in homes, on the street, and in the waters where it bred. Ditches were cut to drain swamps, and brush was trimmed away to make it easier for spray to reach the insects. Men with pumper tanks and nozzles sprayed more than two million rooms in more than 300,000 homes. Italian Air Force veterans were trained to use airplanes and helicopters to lay DDT upon vast fields. Boats made specifically for an antimosquito navy were sent to spread larvicide on rivers and ponds. More than 260,000 kilograms of DDT—roughly 256 tons—were eventually dusted over the island.

The objective was to wipe the mosquito off the island and thereby eliminate malaria forever. Malaria prevalence plummeted to just four cases in 1950 and would eventually disappear. From a public health standpoint, the Sardinian campaign was a great victory.

But the project's commanders were not so pleased with what their experience suggested about DDT and mosquitoes. At the end of the spraying, they had sent scouts out to look for *Anopheles labranchiae*. The searchers had found both larvae and adults, alive and thriving. Five years of work and several million dollars had been invested, and yet the mosquito still lived. As the directors' final report noted, the project "from this point of view, must be considered a failure."

If the survival of *Anopheles labranchiae* in Sardinia surprised public health experts, they were shocked by the news that soon came from another corner of Europe. Not long after the United States and Italy began their island experiment, the Greek government had begun an all-out war on malaria using the same super weapon, DDT.

Greece had a history of malaria outbreaks that predated Hippocrates. Most recently, in 1942, half the country's population had been infected. In 1947, a war of eradication was declared with the goal of freeing Greece once and for all from malaria. The mosquito in question was *Anopheles sacharovi*, a notorious malaria carrier that breeds in brackish water around the Mediterranean. With DDT supplied by the United Nations and with military veterans doing the spraying, the Greeks went after the mosquito with warlike intensity. An American named D. E. Wright, who was a disciple of Gorgas, was put in charge.

As far as most Greeks were concerned, the campaign was a glorious success. Spray crews were housed and fed by local villagers who welcomed them as a liberating army. Pilots in biplanes skimmed twenty feet above the ground to attack swamps with a DDT fog. And where olive farmers were lucky enough to have their groves catch some of the spray, the collateral deaths of destructive caterpillars meant a much bigger haul at harvesttime. In every town, residents were thrilled to discover that flies, fleas, lice, roaches, and other pests disappeared along with their mosquitoes. DDT became so popular that a few pilots took unauthorized turns over neighborhoods where friends and family lived just to kill sand flies.

Except for the accidental deaths of some silkworms and honeybees, the Greek DDT offensive seemed an unblemished success. In 1948, malaria appeared to be essentially gone. Whereas 16 percent of Greek children had previously tested positive for malaria parasites, none could then be found. Then something strange happened. The flies came back. Soon afterward a malariologist eating lunch at a country inn noticed something even more troubling: several of the dreaded *Anopheles sacharovi* mosquitoes flitting about a room that had been treated with DDT.

Finally, in 1951, the men who had just sprayed a village noticed that these mosquitoes had returned in a matter of days and had recommenced biting.

Laboratory investigations demonstrated that *Anopheles sacharovi* could adapt, due to natural selection, to the presence of DDT. Some officials hoped that the Greek mosquitoes were somehow special, and other species might not pull off the same trick, but they were wrong. Resistance was being noted in Lebanon and Saudi Arabia. And then came the worst news of all: massive DDT resistance in *Anopheles albimanus* created by the agricultural use of DDT in cotton fields in El Salvador. In that case, the mosquitoes were adapting even before an impending antimalaria campaign could begin.

The lowly mosquito's ability to resist DDT had political as well as scientific implications. As every student of history knows, the West's cold war hostilities with the Soviet Union began immediately after the end of World War II. As the Communists worked to bring one nation after another into their sphere, the United States and Western Europe responded accordingly.

In the geopolitical competition, new technologies yielded the best propaganda material. Whether it was *Sputnik*, atomic energy, or new vaccines, the competing systems presented every advance as evidence of a superior political and economic system. Whenever possible, breakthroughs were delivered to allies and potential allies in order to extend political influence. At the head of this effort, the United States, through its Agency for International Development (AID), created countless programs designed to use American money and know-how in this way.

DDT was going to be a major weapon against communism

as well as mosquitoes, bringing health to a world that would have to notice that it came from America. The resistance problem threw an unexpected and disturbing variable into the picture—time. Although many mosquitoes ultimately adapted to DDT, liberal and broadscale use of this extraordinary insecticide would disrupt transmission of malaria long enough for the disease to begin to disappear. Other forces might then break the cycle of transmission. This is what happened in Greece where, in the end, the mosquitoes survived but malaria did not.

Eradication soon had a new definition. Instead of using DDT to eradicate the vector, the antimalaria warriors came to see it as a weapon to be mobilized against the parasite. They realized that if the mosquito could be beaten down, the cycle of transmission would be broken long enough for every person whose blood teemed with the pathogen to be cured before it could be passed to anyone else. The parasite would disappear from a local population, and if the mosquito returned, it would be merely a nuisance, not a potential killer.

During the 1930s and 1940s, an antimalaria program had been launched almost spontaneously in the American South with little fanfare. In 1935, malaria infected roughly 130,000 Americans, causing about 4,000 deaths each year. By 1950, however, only isolated cases were noted. The event that became known as the malaria "recession" was finally documented in the postwar era by Alexander Langmuir in the course of creating the Epidemiological Intelligence Service (EIS) of the Centers for Disease Control and Prevention. The EIS developed into an outstanding rapid-response team of shoe-leather epidemiologists that today identifies disease outbreaks of any kind and tries to prevent their spread. (Some other nations have similar agencies.)

Langmuir's first mission was to map the malaria problem in

the United States and to eliminate the disease from within the nation's borders. He began by formulating the rule that active malaria transmission could be documented in a locality only when "two cases are associated in place and time." He then assembled teams of epidemiologists and sent them out to discover where this rule would prove that the disease was entrenched. They returned with a surprising answer—nowhere. Try as they might, Langmuir's investigators couldn't find a single place where an antimalaria program was warranted. He issued a report and with the stroke of a pen declared America malaria free.

In retrospect, it is evident how malaria was quietly defeated in the United States. Probably the largest single factor was the Tennessee Valley Authority, which built dams, power systems, drainage projects, and irrigation networks throughout the parts of the deep South that were most malarious. Every water project was designed to limit mosquito breeding.

Where the system of "water-level fluctuation" (described in the previous chapter) could not be implemented, standing water was monitored for anopheline larvae, and larvicides were applied. Arsenic-based Paris Green was used initially. Various other synthetic chemicals were employed after World War II.

Anopheline larvae are more readily destroyed than those of other kinds of mosquitoes. They breathe through two pores located near the end of their abdomen and keep their bodies in close contact with the surface of the water. They generally feed on material directly associated with the surface film. (Other kinds of larvae hang down from an air tube that separates their bodies from the surface film and feed on material that is suspended in the column or water or adheres to the bottom of the water body.) Anophelines, then, can readily be recognized and destroyed by insecticides that float on or just under the film. *Anopheles quadrimaculatus*, the vector species in the eastern United States, exploits fairly large bodies of stable water. The TVA system of surveillance and response proved to be highly effective.

Beyond direct antimosquito elements, the TVA's electric grid and economic development programs brought new wealth to the region. As farmers began to use machinery and chemicals to increase their yields, they began to climb out of poverty. Shacks were replaced by tidy homes, with screens on windows and doors. As land became more valuable, low-lying areas that once bred mosquitoes were drained and put into production. Wealth created a cycle of health, and improved health helped generate more wealth.

During the brief time I worked for the TVA in the early 1950s, I realized why malaria became overwhelmed by the changing economic landscape. Along with my other duties, I was asked to help determine why the remote Black Swamp region of northeastern Alabama had become malaria free. Numerous cases of malaria had been diagnosed there just a few years earlier, and no specific antimalaria measures had been launched to improve the situation.

The swamp is a vast area of gum trees that grow out of hummocks in the knee-deep water. The water is black with plant debris and teems with snakes and turtles. The farmers we visited were sharecroppers who tilled cotton fields that were carved out of the nearby high ground. Though these families welcomed us, we had to be careful as we traveled the dirt roads in the region because the moonshiners who worked there might not distinguish us from the "revenuers," whom they considered enemies worthy of a charge of buckshot.

We understood why malaria had disappeared as soon as we stepped into the first sharecropper's house. A suspicious-looking white powder was everywhere. The residents readily confessed that they were "borrowing" toxaphene from the cotton operation and dousing their homes with it. They intended to protect themselves from *Coquillettidia perturbans*, a particularly aggressive mosquito whose larvae breathe by attaching themselves to the roots of the cattail plants that grew in the nearby swamp. The toxaphene worked not only against the *Coquillettidia perturbans* but also against local *Anopheles quadrimaculatus*, the malaria vector.

The resourceful sharecroppers of Black Swamp had found their own way to ward off an annoying pest and, in their own informal way, had also broken the malaria cycle for good. Because exposure to toxaphene is harmful to people, alternative antimalaria measures were recommended. But given that the pesticide was acquired for free, and that it had worked so well, this advice may not have been followed.

In an ideal world, the malaria recession that occurred in the Black Swamp and throughout America would have been better understood at the time by science and public health experts. With the benefit of historical perspective, it's clear that certain kinds of economic development, especially improvements in rural communities, had been the keys to a positive cycle of wealth and health. In the end, malaria was beaten without a Soper-like war on the mosquito vector.

The American experience should have told officials in the U.S. State Department and the United Nations, who were gearing up for a global malaria fight, something about the weapons they might choose. Instead, they held on to DDT, deciding to apply more of it, faster, and everywhere that they could. In the end, the world would reach one of two end points. The one that was desired was the total and irreversible eradication of the vector for human malaria. The other would be the ultimate impotence of DDT, due to the development of insecticide resistance.

Seven years was how long it seemed to take for resistance to arise in mosquitoes under attack from DDT and to become an insuperable obstacle to its continued use. If we were to employ this chemical wonder in an all-out attempt to rid the world of malaria, our goal would have to be accomplished within that span of time. This was the conclusion reached in 1956 in a seminal report issued by the International Development Advisory Board (IDAB) to the U.S. State Department.

Guided mainly by Dr. Paul Russell, Soper's colleague and a true believer in eradication, the IDAB laid out a bold plan requiring $520 million in 1955 dollars (the equivalent today would be many billions of dollars) to defeat malaria worldwide. "Eradication is economically practicable today only because of the remarkable effectiveness of DDT and related poisons," the board declared. But because of resistance, it added, in capital letters, "TIME IS OF THE ESSENCE."

The humanitarian reasons for a world war on malaria were obvious, but the report suggested other justifications for America paying the bill for the assault. The first was economic. Malaria was debilitating workers and consumers in countries that trade with the United States. Their lost labor made goods that the United States imported more expensive. Their malaise also retarded development, meaning fewer markets for high-priced American goods. Ultimately, America paid a hidden, $300 million "malaria tax"—in higher prices and lost sales—every year.

But beyond dollars, the board saw potential political benefit in a huge malaria eradication initiative; it would win America friends. Unlike other development programs, which rarely lead to "visible evidence of progress," a malaria eradication project made life better, immediately, for every family it touched. The point was obvious. DDT spray teams that appear twice a year with mosquito-killing agents labeled MADE IN USA would reinforce America's positive image.

As proof of the political benefits of DDT, the board pointed to instability in Iran in 1950. From the vantage point of 1956, it argued that the presence of American-led mosquito killing teams throughout the country made a "concrete manifestation of our sincerity and mutual interest." This contributed materially to the victory for a pro-American regime led by the Shah of Iran. (Many years later, the world would know that America's covert aid to the Shah, through the Central Intelligence Agency, had much more to do with events there than the Iranian public's positive regard for DDT from the United States.)

The IDAB message—that America should immediately fund a program literally to douse the world with DDT and end malaria—could not have surprised anyone who knew the board's scientific adviser, Paul Russell. Just one year before, Russell had published a book with the audacious title *Man's Mastery of Malaria*. In it, he expressed an evangelical zeal for eradication. "The verb 'to master' does not imply an end to the matter," he writes, "rather it suggests that, having prevailed over an opposing force one has assumed moral responsibility for keeping it under control."

Here was the sum of Russell's medical, scientific, and even family background coming to the fore. The son of a New England preacher, Russell had wanted to serve both God and man as a medical missionary. In his work with malaria, he found a true moral purpose. He was excited by DDT, which he called "a wonder," and his faith in eradication was all but unshakable.

The DDT evangelists were persuasive enough to convince the State Department and then two prominent senators—John F. Kennedy and Hubert H. Humphrey—that victory over malaria was attainable. The timing couldn't have been better. In 1957, the Soviet's launch of the *Sputnik* satellite revved America's competitive spirit to an almost paranoid pitch. The government was ready to do anything to prove the might of American science. A project to use an American-manufactured miracle technology (DDT) to rid the world of malaria fit perfectly.

Congress approved the program and the Agency for International Development quickly organized itself to run it. USAID virtually took over the existing World Health Organization program, greatly accelerating and expanding the work under way. President Eisenhower, and later President Kennedy, delivered speeches that declared all-out war on this mosquito-borne disease.

In 1958, DDT was shipped by the ton to dozens of countries that stood on the front lines in the great war against malaria. The commitment to this strategy was total, so much an article of faith in American science that grants for malaria research disappeared almost overnight. If the grand solution—DDT—had been found, then what was the point of research? This became dogma, even at universities. At Harvard, for example, the faculty simply avoided teaching and doing research in malariology. To continue would have been seen as subversive.

The demand for total loyalty that swept the field of tropical medicine may have been a product of the urgency felt by so many at the top. They understood best the time pressure in our gamble. Wherever DDT was applied, it would have to work its magic fast. But there was another factor to fear: recolonization. No country that rid itself of malaria would be safe if its neighbor did not join in the effort. Mosquitoes and travelers could inevitably reintroduce the infection. Here lay one final, terrible problem. A population that has been rendered malaria free for a few years loses its immunity. And if the pathogen returns, everyone will be more vulnerable to severe disease than they were before they received the benefit of this intervention. Immunity works to modulate the symptoms of malaria but does not prevent reinfection.

With the terrible possibilities of DDT resistance and recolonization hanging over their heads, the U.S. Agency for International Development and the World Health Organization worked at a fevered pitch. Training institutes were opened, malaria surveys were conducted, manuals were assembled. By 1960, sixty-six countries were embarked on the prescribed spray campaign. Another seventeen were either planning or negotiating to get involved.

Early success was seen in the few island countries that were selected as showcases for the scheme. Taiwan was literally cleansed of malaria, with the prevalence of infection dropping from 8.6 percent to zero. Zanzibar and Jamaica benefited similarly. Sri Lanka, where thousands died in some years, saw fatalities decline to fewer than a dozen per year. Progress was nearly as good in India, where an enormous effort was mounted. Spray teams were organized and sent out to cover malarious regions that totaled more than 1.2 million square miles.

The American-funded effort in India was publicized with photos of men wielding sprayers and buckets of DDT while riding atop elephants. At the peak of the effort, 150,000 people were employed in the Indian eradication project. In 1961, malaria was so beaten down that it accounted for less than 1 percent of that nation's hospital admissions, compared with 10 percent in the mid-1950s. Malaria cases nationwide dipped to fewer than 150,000. This in a country where, less than a decade before, 75 million people were infected and 800,000 died in a single year.

By 1961, just three years into the offensive, the proponents of eradication were adding up their victories. More than 20 percent of the people once plagued by malaria lived in cleared areas. Wherever near total coverage was achieved, malaria was beaten. And the only resistance the program officers had met was not in the mosquitoes but in the offices of the bureaucrats who managed their funds.

Well before the end of the five-year timetable, Americans were bragging about their role in this effort. At a mosquito control conference in California, one federal official made the claim to glory: "Where else could sixty million pounds of insecticide be mobilized in one single year for this fight overseas?" asked Roy F. Fritz. "Where else could we obtain the number of vehicles and the spray equipment which can withstand eight hours to ten hours a day's use day in and day out, year in and year out? Where else could that be produced? It isn't produced anywhere else."

. . .

In many countries, charismatic leaders arose to become the generals in the war against malaria. None were more aggressive, or successful, than Arnaldo Gabaldon, the round little force of nature who became Venezuela's top mosquito fighter. Gabaldon had seized on the DDT/eradication strategy as soon as World War II ended and instituted one of the first national programs for defeating the disease. (He later managed President Lusinchi's successful election campaign in Venezuela.)

By the time America began financing a worldwide attack on malaria, Gabaldon had already anticipated some of the problems that would arise. Despite achieving near elimination, he had been vexed by small outbreaks caused by two factors: mosquitoes that survived and people who had not allowed their homes to be sprayed. Problems with access to certain homes could be overcome, he reasoned. But the mosquitoes were a more difficult challenge. A number of them had become resistant, he noted, but Venezuela also had a problem with certain *Anopheles* that didn't hang around inside homes long enough to fall victim to indoor spray.

"They leave the dwelling immediately after biting," he observed. And unfortunately this "habit of living outdoors" did not reduce the insects appetite for human blood. "*Anopheles nuñeztovari* in Venezuela. . .," he added, "maintains a human blood preference of around 80 percent, and a man-biting rate of more than 100 during a night indoors." (This means that a sleeping person would receive an average of 100 bites from different mosquitoes per night.)

Gabaldon's experience led him to advocate a public health strategy that was shunted aside in the rush to worldwide eradication. He preferred a sort of "fire brigade" approach that would rely on surveillance as much as chemicals. Monitors would look for a spike in malaria cases, and in response, Gabaldon's brigades would descend with DDT to be used in a local, controlled way. Rejected elsewhere, this was how Gabaldon would keep malaria in check in Venezuela for more than twenty years. Years later, when asked why the grand malaria battle of the cold war era was lost, his face became flushed, he pounded his fist on his desk and declared, "Because they did not listen to me!"

Exactly one year after Roy F. Fritz crowed about American technology and the coming defeat of malaria, the Royal Society of Tropical Medicine and Hygiene met at Manson House, a brick row building in London named for Sir Patrick Manson. These scientists gathered to review the progress, or lack of it, in antimalaria battlefield reports from around the world.

"It appears," noted scientist M. J. Colbourne, "that resistance is seriously interfering with progress in several countries." Though he suggested that a change of insecticides might save the day, there were still other difficulties, including those dangerous "exophilic" species that flitted indoors to bite but refused to rest long enough to be killed by DDT residue. In the end, said Colbourne, "the achievement of worldwide, or even continent-wide eradication does not appear practicable in the near future."

As the veteran British mosquito fighters continued to discuss the problem, it seemed to grow more complex. Sir Gordon Covell, who spent fifteen years fighting malaria mainly in India, noted the threat posed by the mosquitoes in Pakistan, a country which had not embarked on a DDT program. Many countries, especially in Africa, would face problems as trade and proximity to infected areas threatened to reintroduce malaria. Covell also wondered, aloud, if DDT's major successes may have depended on the fact that the mosquitoes that were attacked were the house-bound type, easily killed by indoor spray. In fact, all of the vector mosquitoes in places such as Greece, Sardinia, and Taiwan were these easy-to-kill kinds.

Another British expert, John McArthur, feared that many techniques that worked in specific locales would be forgotten in the rush to apply DDT to every malaria problem. He recalled fighting a kind of mosquito in Borneo that he discovered breeding only in shady jungle. "With quite limited clearing of jungle," the vector was pushed away from settlements. Soon, where 95 percent of the population once experienced splenic enlargement due to the malaria parasite, the rate dropped to 45 percent. But when McArthur proposed an experiment comparing this strategy with one that employed DDT he was rebuffed by officials who had become enthralled with the pesticide.

"Since then," noted McArthur, "reports on the effectiveness of spraying in *Anopheles dirus* areas throughout Southeast Asia are conflicting. Some claim that spraying has succeeded. Others claim that it has failed."

Failure was going to be noted, soon, all over the globe after the seven-year time span and large-scale American funding expired. Taiwan held to its eradication record, but Sri Lanka began to see the beginning of a malaria comeback that would produce half a million cases in 1969. (In this same year, the World Health Organization officially recognized the failure of eradication.)

In other countries the story was to be the same—the gradual appearance of resistant mosquitoes and the subsequent emergence of malaria. In Indonesia, malaria increased fourfold between 1965 and 1968. India held on longer than most, but a massive epidemic announced malaria's return in 1976 when an estimated twenty-five million people were stricken. Due to the malaria eradication effort, many people had lost all immunity to the parasite: their new infections were particularly dangerous.

In the early 1960s, those who pushed the military-style assault on mosquitoes with DDT clung to their strategy. If only a little more time and DDT were applied, they argued, the desired result would occur. But the money, the energy, and, most important, faith in this solution had been running out. The final blow, at least as far as public opinion in the developed world was concerned, came in the form of a book by a middle-aged biologist who once showed her students the strange mating behavior of polychete sea worms near the dock at the famous Woods Hole Marine Biological Laboratory.

Published in 1962, Rachel Carson's *Silent Spring* took on a host of nascent ecological issues, including problems associated with radiation, but its major effect was to challenge the widespread assertion that DDT was safe. Beginning with convincing evidence linking the pesticide to the decline of certain bird populations, Carson went on to describe health problems in workers who handled the chemical and liver cancers in exposed fish. She argued for restraint.

It was from Carson that the general public learned that DDT was found in mother's milk and could accumulate in the bodies of their babies. And it was Carson who told the public that as early as 1950, the federal Food and Drug Administration warned, "It's extremely likely that the potential hazard of DDT has been underestimated."

*Silent Spring* was one of the first popular works to bring environmental concerns to a broad audience. It also revealed that science was not a monolith. Indeed, in one passage after another,

Carson presented the work of respected researchers who challenged the DDT establishment. They described how in some cases DDT spraying killed other animals and allowed for an explosion in the abundance of certain pests. She also explained in clear terms how insecticide resistance developed.

"Spraying kills off the weaklings," she wrote. "The only survivors are insects that have some inherent quality that allows them to escape harm. These are the parents of the new generation, which, by simple inheritance, possess all the qualities of "toughness inherent in its forebears." Under Carson's guidance, it's pretty easy for any reader to understand how insect populations that give rise to new generations in a matter of days can quickly defy a pesticide.

As convincing as her science may have been, it was Carson's gift with prose that was most persuasive. "The 'control of nature' is a phrase conceived in arrogance, born of the Neanderthal age of biology and philosophy, when it was supposed that nature exists for the convenience of man," she concluded. "The concepts and practices in applied entomology for the most part date from the Stone Age of science. It is our alarming misfortune that so primitive a science has armed itself with the most modern and terrible weapons, and that in turning them against the insects it has also turned them against the earth."

*Silent Spring* was published eight years before the creation of the U.S. Environmental Protection Agency. But almost immediately, several states began reviewing the use of pesticides, and by 1968, half a dozen had banned particular chemicals. Four different national research commissions conducted studies on DDT and recommended that its use be phased out. A year later, the United States Department of Agriculture began banning its use on certain crops. In 1971, the EPA held hearings on DDT that resulted in nine thousand pages of testimony. On December 31, 1972, the agency issued a press release announcing that "the general use of DDT will no longer be legal in the United States after today."

DDT's supporters continue to argue that, despite the application of more than 1.3 billion pounds of the chemical in the United States, not one human death can clearly be attributed to its use. But this argument could not stand against the perceived long-term danger. And with the chemical banned in America, the prospect for its use overseas dimmed as well. How many leaders abroad would be able to assure citizens that a pesticide that Americans no longer produced, and feared using themselves, was actually safe?

Nevertheless, a worldwide ban on DDT would be a severe loss for public health workers because the pesticide has properties that render it an invaluable tool against malaria. When used for this purpose, it is applied close to where people sleep, on the inside walls of houses. After biting, the mosquitoes generally fly to the nearest vertical surface and remain standing there for about an hour, anus down, while they drain the water from their gut contents and excrete it in a copious, pink-tinged stream. If the surfaces the mosquitoes repair to are coated by a poison that is soluble in the wax that covers all insects' bodies, the mosquitoes will acquire a lethal dose.

No chemical compares to DDT as a weapon against the resting mosquito. First, it is potent. Just two grams of DDT per square meter of wall surface is more than enough to kill a mosquito within its usual one-hour resting period. Second, it is inexpensive. It is also easily stored and transported, and relatively safe for the person doing the spraying. Best of all, it remains effective for many, many months.

The total ban on DDT's use in the United States deprived American public health officials of a weapon that could have been safely used. Even today, when there are many chemicals available to kill mosquitoes, DDT retains many advantages. It is the ideal insecticide of first use. This is because the resistance that mosquitoes develop after being exposed to DDT does little to protect them against the other, more expensive insecticides that wait on the sidelines. However, mosquitoes hit first with one of those other compounds—such as malathion, sevin, or permethrin—develop a broader resistance that partially protects them from DDT as well. A spray program based on the use of chemicals in any of these alternatives also tends to be about three times as expensive as one based on DDT. When used correctly and with restraint, DDT appears to be irreplaceable in antimalaria programs.

In the year 2000, DDT was nearly outlawed worldwide under the terms of a United Nations Environmental Program treaty. It was to be classified as one of the unsafe "dirty dozen" of the Persistent Organic Pesticides, known as POPs. In December 2000, however, a treaty conference held in South Africa agreed to a "dirty eleven." DDT was excluded from proscription. The chemical is now manufactured only in China and India, and it is to remain available solely for use in antimalaria programs. This most recent battle over DDT's status was intense and the outcome crucial for helping to protect human health around the world.

By the time that the postwar effort to eradicate malaria waned, many of the important vector species had developed some degree of resistance. Where malaria workers turned to substitute insecticides, some mosquitoes learned to resist those too. At the same time, the malaria parasite itself began to demonstrate its ability to evolve resistance to our best medicines, and drug resistance accelerated as these medicines became more readily available. In too many poor households people obtain the medicine, use just enough to ease their symptoms, and then hoard the remainder for the next wave of illness. Malaria parasites frequently are exposed to sublethal doses of drug. In this manner, malaria sufferers turn their bodies into ideal breeding sites for drug-resistant parasites.

Drug-resistant malaria was encountered by the U.S. Army when it was mired in the Vietnam War. Through much of the conflict, more men were knocked off duty by the malaria parasite than by war wounds. Most troubling was the alarming frequency of cases that proved to be resistant to the most widely used drug, chloroquine.

Fortunately, the U.S. Army had been conducting an active program of drug research, screening thousands of compounds for antimalaria activity, and had developed a small stable of alternative compounds, including mefloquine (also known as Larium) and halofantrine. These drugs destroy the parasite's ability to feed on red blood cells and the parasite essentially starves. The drugs can be used as prophylactics, and are commonly taken by people traveling to malarious areas. One must be careful with them, however, because they can cause vertigo and disturbing psychological side effects, including paranoia.

In spite of its almost unbearable side effects—ringing in the ears, fever, malaise—the old standby, quinine, remained available. Primaquine was developed to kill parasites in the liver. And a sulfa-containing drug, Fansidar, also proved to be effective. (More recently artesunate, a derivative of a Chinese herb known as qing hao shu, became available.)

One or another of these drugs saved the lives of virtually all of the U.S. soldiers who became infected in Vietnam. By the 1990s, however, malaria parasites came to tolerate exposure to these drugs too, a troubling trend that has become particularly evident in Southeast Asia.

In the years after the Vietnam War, drug-resistant malaria crept across Asia, into Africa, and then invaded South America. Having invested roughly $1 billion in 1970 dollars in eradication, various antimalaria agencies refused to give up. They turned to the notion that a vaccine might prevail where mosquito-directed efforts had failed.

Here again, the complicated nature of malaria continues to temper human ambition. Malaria parasites have as many as seven thousand genes. And whereas other microbes generally present the same face to their hosts, malaria undergoes many distinct developmental stages in both its mosquito and human hosts. The surface properties of each stage are unique. A vaccine designed to thwart one developmental stage is useless against any other.

Many other fundamental problems stand in the way of malaria vaccine development. One of the more obvious is the difficulty that any researcher experiences in translating experimental results in animals into observations applicable to human beings. Although many creatures experience malaria infection, each generally is host to its own specific kind of parasite. Human malaria parasites do not infect birds or lizards. The malaria parasite that will infect a rat won't colonize a human. And not all nonhuman primates are susceptible to human malarias and vice versa. Even if this difficulty could be overcome, science faces a further challenge in the endless varieties of human malaria parasites. A man receiving a vaccine developed against his hometown brand might remain susceptible to whatever strain afflicted a village two hundred miles away and probably in his own village some time later.

An intense and quite notorious vaccine-development program was conducted by AID during the 1970s and 1980s, when the agency devoted tens of millions of dollars to vaccine research. (In one three-year period alone the total was more than $20 million.) Despite repeated failures and little evidence of progress, grants and contracts continued to be awarded, and AID often reported that one project or another was at the point of a breakthrough. In the mid-1980s, the agency issued a press release that predicted that a workable vaccine would be available before 1990. None was forthcoming. To add to the scientific injury, the last days of the AID malaria vaccine program were marred by criminal scandals. Researchers and contractors were indicted. James Erickson, who managed the project, pleaded guilty to criminal charges. One man hired to supply monkeys was caught trying to smuggle them across national borders but escaped in his private plane. A Hawaiian investigator and his accountant were convicted of misusing AID funds. In the end, hopes were inflated, and the only people who benefited were those who pocketed AID money. The hundreds of millions of people who live with the daily threat of malaria received little from this unfortunate episode.

Despite the obstacles and failures, many research institutions remain committed to malaria vaccine research. Most prominent is the effort housed in the U.S. Navy Medical Research Institute and led by Steve Hoffman. Highly sophisticated techniques in vaccinology are applied, including injection of "naked DNA," parasite genes that generate the parasites' own proteins within the body of the vaccinated person. An immune response is stimulated. Although this and other efforts continue, the only effective vaccine so far developed must be delivered via the bites of at least a thousand irradiated, malaria-infected mosquitoes. (The radiation renders the parasites unable to multiply.) The protection conferred by this experimental vaccine, however, is only transient.

The dream of a true vaccine remains elusive. In recent years hope has been focused on the notion of stimulating private industry to develop an effective malaria serum. Economist Jeff Sachs of Harvard is proposing the offer of a vaccine purchase fund that would guarantee a market for the company that produces a useful product. Such a guarantee might yield sales worth a billion dollars annually.

No history of the monumental malaria eradication crusade would be complete without acknowledging several key points. Although malaria continues to destroy human lives and impede happiness, the recent eradication campaign preserved human health and lives, at least temporarily, around the globe. It's reasonable to suppose that many millions of people escaped death because they got through their vulnerable childhood years without infection. Sadly, many of their children and grandchildren will not now be spared a malarious fate, as this infection continues to stage its raging comeback.

It is also necessary to acknowledge America's own major failure to destroy a certain dangerous mosquito species. Under pressure from Latin American countries that had run the yellow fever vector out of their territories, the United States promised to eradicate *Aedes aegypti* from the Americas in the mid-1960s.

There was more than a little irony in this situation. After all, America had gone to war with Spain, in part, because of the danger of yellow fever spreading from Cuba into nearby lands. Now the same Latin American countries that had been blamed for sending pests north were quite understandably demanding protection from the same mosquito harbored in the United States.

Mainly a container-breeding insect, *Aedes aegypti* would be attacked by DDT applications and by crews that located and removed disused tires, jars, and other man-made containers that collected water and supported breeding. Thousands of workers were trained for the task. Fleets of trucks were equipped with sprayers. Nine states, from Texas to South Carolina, were notified that they harbored the mosquito and would be battlegrounds in the war against it. A determined government notified laboratories that used the mosquito for research that these colonies would be destroyed. This "guinea pig" of the medical entomology laboratory was to be lost, even in northern cities where these insects could not survive the winter.

Once the effort began in earnest, so did the opposition. Laboratories successfully fought the order to destroy the mosquito colonies that they used for experimental studies. And in one community after another, the U.S. Public Health Service encountered residents who didn't want anyone traipsing through their backyards looking for old tires or blanketing their neighborhoods with DDT. They had read (or heard of) *Silent Spring* and would have none of it.

As legal costs mounted, the eradication effort slowed. Then the EPA banned DDT. The war was over. Latin America would continue to live under the threat that the yellow fever vector could return via the United States. It was another victory for the mosquitoes. (This fact became clear to me on the day I visited the *Aedes aegypti* eradication program's headquarters near Atlanta and found the little devils breeding in a can that had been discarded near the agency's parking lot.)

When the great war on mosquitoes was begun, Fred Soper and Paul Russell declared victory for their theories of eradication as a solution to malaria. Generations of experts had argued over whether eradication was really feasible, and for many years those who favored more cautious "control" or "suppression" programs had prevailed. But with the U.S. government's embrace of the eradication concept, the matter seemed completely settled. It became such an article of faith that students who hoped to enter the field were warned that they had joined a dying profession. The vector mosquito was soon to be wiped off the face of the earth.

As failures mounted, however, and it became clear that worldwide eradication would never be achieved, Soper and Russell quite literally ceased discussing the matter in scientific forums. Russell, who had accepted a position at Harvard in 1959, demonstrated his disappointment by avoiding the subject altogether and steadily withdrawing from contact with students and faculty. By 1968, he had retired to an isolated village in Maine. Soper remained a heroic figure because of his work in Brazil, but he distanced himself from the global malaria initiative. Remarkably, his 1977 memoir, *Ventures in World Health*, makes no significant mention of this grand attempt to apply his theory.

One year after the publication of Soper's book, I would visit Sri Lanka to investigate the resurgence of malaria. The island nation had been on the front lines of the U.S.-funded, World Health Organization–sponsored eradication program designed by Paul Russell. Sri Lanka had been considered to be the ideal place to employ Russell's DDT-based strategy. As a relatively small island, the country was easier to cover with DDT, and the comings and goings of people and pathogens were easier to monitor.

Besides geography, Sri Lanka offered a well-educated population, and a society that seemed more prepared than others for a national, cooperative effort. Buddhism and Hinduism exert the most powerful influence on Sri Lankan culture, and they fostered deep respect for community. It was not difficult to persuade Sri Lankans to allow their homes to be sprayed with insecticide for the good of all.

Finally, the Sri Lankans were deeply motivated by a sad history. For centuries, periodic malaria epidemics—typically three to five years apart—had swept the island with devastating effect. Many Sri Lankans had vivid memories of the terrible outbreak in 1935 in which more than twenty thousand people died.

Given their civic-mindedness and their respect for the malaria problem, millions of Sri Lankans had joined in Russell's campaign. Every household opened its doors to the DDT spraying teams. Victory over *Anopheles culicifaces* seemed near in 1963 when the malaria count dropped to a mere eighteen human cases. It was logical to assume that the zero point would soon be reached. This apparent success came just in time, because American funding for the eradication program was based on the knowledge that DDT resistance would likely develop in mosquitoes after five years. Five years, and the funding, ended in 1963.

Malaria began to make its comeback almost as soon as the last spray can was put away. By 1969, the death toll exceeded 220, and infections topped 520,000. In 1977, the overall number of cases declined, but their severity worsened as a more virulent parasite—*Plasmodium falciparum*—spread. In all, 501 Sri Lankans died of malaria that year, a number greater than the death rate *before* the eradication program began.

Besides bringing illness and death, the resurgence of malaria in Sri Lanka worried officials of the World Bank, who were about to fund a massive project to dam and channel the Mahaweli River. The Mahaweli runs from the mountains of central Sri Lanka north to the Bay of Bengal. The project would provide electric power and irrigation for vast tracts of farmland. But big water projects can mean big mosquito problems. The World Bank therefore assigned me to assess the risk associated with the Mahaweli scheme and, tangentially, to unravel the mystery of malaria's return.

My first stop was the capital city of Colombo, which squats beside a harbor on the west coast of the island of Sri Lanka, just seven degrees north of the equator. One hundred miles to the west, across the Gulf of Mannar, lies Cape Comorin, the southernmost point of the Indian subcontinent. To the south, the warm Indian Ocean stretches seemingly to infinity.

In the rainy season, successive low-pressure systems gather up the moisture of the sea and drag it heavily onshore. In 1978, air-conditioning was still rare in Colombo, so indoors and out, everything and everyone was hot and moist. In the morning, my clothes became soaked with sweat before breakfast. At night the sheets on the bed were already soggy when they were turned back.

In the capital, I visited so-called fever clinics, which treated some of the tens of thousands of Sri Lankans who suffered from malaria. The relentless rhythm of the heat and monsoon rains echoed the steady tempo of the fevers and chills that surged in their bodies.

The puzzle of malaria's resurgence was made more complex by the diversity of Sri Lanka's ecosystems. A coastal plateau encircles the island. In the north, a semiarid plain supports rice paddies and vegetable farms. Jungle highlands dominate the central and southern regions. This mountainous area includes Mount Adam, where Islamic folklore has it that Adam and Eve sought refuge after they were cast out of Eden. The tropical forest offers a sample of the wilderness that once covered the entire island. The main city in this region is Kandy, which was the seat of the last kingdom to capitulate to European invaders in the nineteenth century.

Mosquitoes could not, on their own, return malaria to Sri Lanka. A human host was needed, and the human population *should* have been protected inside houses that were periodically painted with insecticide. This had been done in a very careful and systematic way. All homes were numbered by the authorities, to help indicate when they had been treated. The numbers could later guide crews when reapplication was necessary. Here again, the countryside gave the answer. Miles away from the large cities we came upon crews of workers who had been brought into the region to build highways. They lived in tiny tentlike shacks (made of palm fronds) that could not be sprayed or numbered. These men and women were more than likely infected and then spread the disease whenever they camped near towns.

Other subsistence workers, who had no fixed abodes and therefore were not accounted for in Russell's grand plan of attack, made similar contributions to the return of malaria to Sri

Lanka. Practitioners of slash and burn farming—called chena farmers—lived in huts in jungle clearings far from the roads. Gem miners lived in similar informal housing with no protection from the bites of mosquitoes. The miners dug in the gravel and then used sieves to search for emeralds and sapphires. They left behind thousands of holes filled with clean, clear water. *Anopheles culicifaces* will breed only in clean, clear water. The miners' holes were perfect for these mosquitoes.

It was apparent also that once it was constructed the Mahaweli River project could make malaria much worse. One element of the plan called for the extension of many miles of irrigation canals to flood rice paddies. The problem wasn't the new paddies. The problem was the canals. Filled with clean, clear water, they could create mosquito hatcheries many miles long whenever they leaked onto the adjacent farmland.

Expeditions across Sri Lanka had helped me to explain how mosquitoes had been able to reestablish the malaria vector, and gave us an indication of how the Mahaweli scheme could make things worse. But the trekking did not fully explain why the eradication campaign of the 1960s had failed. By all accounts, the DDT that had been sprayed in Sri Lanka had been so effective that the defeat of the mosquito should have been complete.

The mystery would have remained but for my review of the government's records from the malaria campaign. The files were kept in a small office down a long corridor in the dilapidated Ministry of Health building in Colombo. Thick manila folders contained the plan of attack, the records of fieldwork performed, and surveillance results. Everything was there, right down to the listing of supplies and their use.

Amid the arcane and mundane lay a brief document that revealed the hand of humankind in the return of malaria to the island. It was a memo that decreed that the eradication team change its criteria for reporting the rate of the mosquitoes' resistance to insecticide. In the beginning, the effectiveness of DDT was confirmed when 100 percent of the test mosquitoes were killed within an hour. As long as this kill rate was observed, the pesticide could be deemed effective and spraying could continue.

But halfway through the program, the time limit was suddenly doubled, to two hours. Though the reason was not recorded, it was obvious that some mosquitoes were developing resistance and the change was made to justify continued spraying. It also allowed the team to reassure political leaders and the public that their efforts were succeeding. In fact, they had already failed and the mosquitoes held the upper hand.

By changing the criteria for measuring resistance, officials in Sri Lanka gave the appearance of success where it did not exist. It was, in the end, the kind of deception that people practice when they are trying their very hardest to do something grand, even heroic, in the face of a terrible problem. It was a deceit born of hope and the deepest desire to believe what *might* be true, because it was so much better than what was real.

In reality, mosquitoes are a pest and a threat that require people to mount a consistent, sophisticated, and even strategic defense. The impulse to smash the enemy must be measured against the knowledge that, in the case of a weapon like DDT, it is possible to go too far.

The law of diminishing returns applies as well to the drugs used to ease the mosquito victim's malarial suffering. Many of these medicines are descended from quinine, the tree bark remedy that Jesuit missionaries borrowed from South American Indians in the seventeenth century. Unfortunately, the more any one of these drugs is used, the more likely it is to encourage the development of resistant strains of the parasite. Eventually, the local vector population will be dominated by resistant organisms.

In Sri Lanka's malaria wards compassion set the stage for the horrible flowering of parasites that would be immune to the standard treatments. Too many patients were being provided with free medicine. Too many were going home and failing to finish the full course of drugs. In their ignorance, a more intractable form of malaria was being bred right in their bloodstreams.

Today a map shaded to illustrate the worldwide distribution of malaria does not look much different from one drawn in 1955, before the great mosquito crusade. A few island nations, most notably Taiwan and Jamaica, have joined the nations of the Northern Hemisphere where malaria has disappeared. But across much of the tropical part of the globe, the parasite predominates. Today's map looks even more ominous when it is marked to indicate where drug-resistant parasites—entirely absent from the world scene in 1955—now roam. Chloroquine can't destroy these parasites, and even our modern multidrug cocktails are sometimes ineffective. Almost every country in sub-Saharan Africa suffers from this more deadly kind of malaria, along with half of South America and much of Asia from Afghanistan to New Guinea. Billions of people are at risk. Every year, 10 percent of the world's population suffers from malaria. Every twelve seconds a malaria-infected child dies.

# 8

# DISEASE WITHOUT BORDERS

Entomologist Mike Nathan was eating dinner outdoors at a Chinese restaurant on Barbados when he spotted some strange mosquitoes trying to feed on his wife. He gently swatted a couple of them, taking care not to crush them, folded them in a napkin, and put them in his pocket. The next day at his lab, he identified the interlopers as *Anopheles aquasalis*, a dangerous vector of malaria. At the time, Barbados was considered free of malaria vectors.

*Anopheles aquasalis*'s role in spreading malaria was defined in a very roundabout way. In the midtwentieth century, when much of the world employed DDT against malaria, officials in what was then British Guiana, three hundred miles south of Barbados, dutifully attacked their important vector, *Anopheles darlingi*.

An aggressive biter that feeds almost exclusively on humans, *Anopheles darlingi* is perhaps the most dangerous vector in the Americas. But because it so favors human blood and rests after feeding indoors, it could be attacked in people's homes. After feeding, this mosquito rests on a wall or the furniture where it would readily be killed by residual DDT.

In a short time, Guyanese spray teams virtually eliminated the vector and malaria began to recede. As one would expect, trade and development improved, and this small territory on the northeast coast of South America began to blossom economically. Then a strange thing happened. As Guyana became a bit more modern, malaria staged a comeback. And because most people had lost their immunity during the disease's recession, a public health disaster loomed.

George Giglioli, a noted Italian malariologist who had worked for many years in Guyana, investigated malaria's rebound but did not find that *Anopheles darlingi* had returned. Instead, he pinned the blame on a mosquito that previously played no role as a malaria vector because it only rarely bit people. *Anopheles aquasalis* had suddenly acquired a taste for human blood. Because she didn't hang around houses like *Anopheles darlingi*, she would be immune to conventional antimalaria measures. Man-made breeding places like ditches and blocked culverts would have to be attacked, along with the upper reaches of salt marshes.

Though the new vector's role in malaria was proven conclusively, mystery surrounded her shift from animal blood to human. Giglioli provided the answer. He discovered that in the economic boom that followed malaria's retreat, the Guyanese had acquired tractors, trucks, cars, and buses, which replaced draft animals such as horses, donkeys, and oxen. The decline in the abundance of these animals had deprived the mosquitoes of their source of blood. In its "desperation," *Anopheles aquasalis* had switched to human blood. With the puzzle of the new vector solved, authorities quickly counterattacked, and malaria was beaten back again. But because *Anopheles aquasalis* can survive nicely in the wild without human blood, it remains abundant in the environment.

In Barbados in the early 1990s, *Anopheles aquasalis* was an unwelcome arrival. Barbadian health officials were mildly concerned, but it was the minister of tourism who recognized them as a potential portent of doom. For years, Barbados had advertised itself as a malaria-free tropical paradise. This was an important selling point for tour operators and travel agents who sold winter vacations to wealthy North American and European customers.

It is difficult to imagine a stronger example of the link between health and wealth. Tourism accounts for nearly 20 percent of the gross domestic product of Barbados. But many foreigners won't travel to places where there is any possibility that they might acquire malaria. For a view of what happens when a resort develops a reputation for disease, one might visit beautiful Buenaventura on the Colombian coast. Perfectly suited to be a major tropical resort, the city has long attracted Colombian tourists. Because of mosquito-borne disease, few foreigners viewed this region as a tourist destination, even before security concerns rendered the country unattractive for other reasons.

Besides the contemporary fear of what malaria might do to a tourist economy, Barbados had a terrifying history of experience with mosquitoes. In the 1600s and 1700s, yellow fever exacted a deadly toll on the island. Men who came seeking their fortunes in sugar knew from the moment they landed that they were in a race against disease. This reality no doubt contributed to the island's reputation as a dangerous place where fortune and death competed on even terms.

The carefully tended image of modern Barbados—a mix of British culture and Caribbean climate—was worth hundreds of millions of dollars to the country's 250,000 citizens. When *Anopheles aquasalis* was discovered, the Pan American Health Organization was called in to investigate. (Founded by Fred Soper, PAHO continues to do health science throughout the Western Hemisphere, with special emphasis on mosquito-related disease.) The government of Barbados hoped PAHO would show them how to maintain their malaria-free claim. PAHO asked me to conduct the investigation.

At the restaurant where Mike Nathan first saw *Anopheles aquasalis*, we saw no evidence of breeding. But just a couple hundred yards away lay Graeme Hall Swamp, one of the few wet places in the island's ecosystem. Most of Barbados is relatively dry. The island has no rivers of any size, and almost every square acre has, at one time or another, been cleared and used for agriculture.

Located about eight miles south of Bridgetown, Graeme Hall Swamp was perhaps the only wild spot left on the island that was moist enough to support mosquitoes, and sure enough we found a small colony of anophelines there. The bad news was that those few mosquitoes in the Chinese restaurant were not lonesome strays. The good news was that they represented a minute population that depended entirely on this small swamp for their breeding sites. The infestation was determined to be "endangered."

Health and environment officials understood that simply attacking the colony in Graeme Hall Swamp carried the risk that they would destroy some of the fauna in one of the rare wetlands on the island. Environmentalists would surely raise a ruckus, and the publicity that would follow a fight over this potential malaria vector would surely damage the tourist industry.

What was the answer?

I recommended doing nothing. The fact was, these mosquitoes were so few in number and their habitat was so limited that few people would ever come into contact with them. What were the odds that a person infected with a malaria parasite would come to Barbados and that a stable chain of transmission would result? Human-mosquito contact was so tenuous that such an event would be a virtual impossibility.

In the end, the government of Barbados reported to the world that a scientific investigation had shown that there was no malaria on the island and that it remained a safe place for the world to visit. Development on the island has continued apace, and tourism now accounts for more than half the economy's foreign exchange income.

The story of *Anopheles aquasalis* in both Barbados and in Guyana illustrates the complexity and the unpredictability of humankind's relationship with mosquitoes and disease in our times. In Guyana, human activity and modern development allowed for a previously harmless insect to acquire fangs, if you will. In Barbados, health and tourism officials were confronted by a small pest population that threatened to disrupt the man-made order that had kept malaria down.

The amazing adapative qualities of mosquitoes and pathogens combine with travel, trade, and natural events to present modern society with endless possibilities for diseases and vector mosquitoes to arise in surprising places. In tropical countries like Guyana, who knows whether the next bit of forest that is felled harbors a heretofore unknown microbe or a new vector mosquito? Industrial nations, where malaria is no longer endemic, have long been accustomed to travelers bringing the parasite and sparking minor outbreaks now and then. But what happens when a new vector mosquito, or pathogen, arrives and becomes established?

This is precisely what happened in greater New York in 1999, when a rash of human encephalitis cases occurred in the borough of Queens. Thus began an episode that would become the most publicized outbreak of a new mosquito-borne disease in history. In the end, the New York experience would become the perfect illustration of the challenges we face in an era of disease without borders.

The first person to be hospitalized in the New York outbreak was a sixty-year-old man who had initially felt mild symptoms similar to the flu—fever, coughing, weakness—but then found himself suffering more severe headaches and partial paralysis. He was diagnosed with encephalitis—an infection of the brain—of unknown origin. Many viruses and bacteria can cause encephalitis, and in New York it was most unlikely, at the time, that a mosquito could have been to blame. A few days after this first admission to Flushing Hospital Medical Center, an eighty-year-old arrived with similar complaints. This man got worse, fast, and suddenly died of heart failure.

Dr. Deborah S. Asnis, who handled both of the early encephalitis cases, notified Dr. Marcelle Layton of the city health department. Asnis understood that encephalitis is rare and oftentimes contagious, and for these reasons public health authorities generally track them. When a third and then a fourth case were documented in the same small area, Layton officially declared it an outbreak. Workers were sent to interview the surviving patients and their families. Blood samples were rushed to the federal Centers for Disease Control and Prevention for analysis.

As this work was done, more people came down with the same symptoms. Eventually the outbreak would include sixty-two cases, more than six times the number diagnosed citywide in a typical year in New York. The elderly seemed especially susceptible. They developed more severe fevers, headaches, and paralysis, and they required more aggressive care. Even with this effort, six more would die.

In the first stages of any investigation of a mysterious outbreak, health department workers look for factors connecting the cases. In this instance, the victims had not eaten in the same restaurant or visited the same places. However, many of the patients had spent time outdoors. Two of the men had chosen to sleep outdoors because their restlessness bothered their wives. New York was enduring one of the hottest summers on record, and people were doing whatever they could to find relief. Many people had taken to spending the hours after sunset on porches, decks, and front stoops.

The single common factor in the outbreak—evenings spent outdoors—led Dr. Layton and her colleagues to consider something in the environment. And the most obvious suspect had to be an encephalitis virus spread by mosquitoes. It could be eastern equine encephalitis, but Triple E wasn't likely to produce so many cases in a densely urban area. No, the logical suspect was a virus that had ravaged the Midwest and the South but had never been seen anywhere near New York City—St. Louis encephalitis (SLE).

. . .

As New York City officials rushed to find an answer to the human encephalitis outbreak, wildlife pathologists in upstate Delmar, New York, were confronted with the carcasses of hundreds of dead birds that had been shipped to them for examination from Long Island and the city. Mainly crows and sparrows, the dead also included owls, hawks, and blue jays. Some of the birds had been observed just prior to their deaths. They stumbled and even flopped over, unable to maintain balance or control their limbs. Autopsies revealed neurological damage, and more complex analyses of blood and tissue were begun.

A similar illness struck horses in nearby Long Island. Veterinarian John Andresen had seen several previously healthy horses stumbling around as if they had suffered some neurological insult. In thirty days, Andresen saw fifteen of these horses, and none of the bloodwork or exams produced a hint of what might be wrong. He began calling every vet he knew, and quite a few he didn't know, but none could help him figure out what was happening.

In the very same time period, birds at the Bronx Zoo were dying in unexpected numbers. Tracey McNamara, the zoo's pathologist, had autopsied eight species including flamingos, owls, and a bald eagle. All had shown signs of encephalitis, but McNamara couldn't identify the cause. As far as she knew, no single disease that was endemic to New York could kill so many different kinds of birds in this way.

Well aware of the ways that animal deaths can signal human illness, McNamara made sure to contact federal authorities, including the Centers for Disease Control and Prevention. She would later recall that the reception she received from CDC scientists who were already consumed with the human outbreak was not quite welcoming. A serious woman with twenty years in the field, McNamara grew impatient trying to get answers from the feds. "Let's say the secretaries could recognize my voice," she would tell the *New York Times*.

Undeterred, McNamara sent lab samples from the birds to the CDC and to labs operated by the Department of Agriculture and the U.S. Army Medical Research Institute of Infectious Diseases. The army would send a portable biohazard containment laboratory to the Bronx Zoo. In the event that a terrible new pathogen was present, this negative-pressure setup would help McNamara and her staff work safely on dead birds.

Dead birds. Dead horses. Dead people. Although she could allow that New York might be under siege from St. Louis virus, or Triple E, McNamara had trouble reconciling the variety of species succumbing all around her. Neither of these common North American viruses was known to cause so much death and illness in the animal world and in the human population. If this was an SLE or Triple E event, it was the strangest one ever seen.

Early in September, Mayor Rudolph Giuliani went to the Whitestone section of Queens, where he stood on a street corner to make a dramatic announcement to the press. The Centers for Disease Control and Prevention had confirmed that New York City was under attack by *Culex pipiens* mosquitoes carrying St. Louis encephalitis virus. Though most people who contract the disease get better, there is no cure, and in previous outbreaks as many as 20 percent of the people afflicted died.

The city had already handed the crisis over to its emergency management teams. Pesticides would be sprayed to kill the mosquitoes, and the health department had begun to distribute leaflets on keeping safe and also samples of insect repellent. A hot line was established to take calls from the worried public. "We will do everything that we can to wipe out the mosquito," promised the mayor.

Standing beside the mayor were health officials who tried to balance the alarm that was being sounded with some hard science about the severity of the outbreak. The odds were stacked in the individual New Yorker's favor, they said. After all, only one mosquito in three hundred is likely to carry the virus. And you could avoid getting bitten by staying indoors and using repellent. But in the unlikely event that you were bitten by that one-in-three-hundred mosquito, you probably wouldn't get very sick. Most people who are infected with St. Louis encephalitis feel nothing more than some aches and fever before their immune systems rid them of the virus.

It's not likely that Mayor Giuliani understood the complexities of immune response to viruses. Nor is it likely that he knew the fine details about past St. Louis encephalitis outbreaks. If he had, he might have been a little less definitive in his announcement in Whitestone. In the decades since the virus was first identified, public health experts had experienced numerous outbreaks. And what was happening in New York did not exactly fit the SLE pattern.

In the past, SLE outbreaks had been signaled by high levels of virus in certain birds. But though infection rates may run high, SLE does not usually kill huge numbers of birds. But in New York in 1999, many crows but also starlings, sparrows, and other birds were dying all over the place. Similarly, in no other SLE outbreak in an American city had exotic birds died at a zoo. But this was happening in New York. And finally, SLE does not usually infect horses and bring them down as veterinarian John Andresen had seen on Long Island.

As they considered the findings of the CDC, many of the experts in New York understood that something was askew. They continued to press the federal laboratories, suggesting that more tests be done. In the meantime, the mayor and his administration declared war on mosquitoes. Squadrons of helicopters blanketed sections of the city with pesticides. Armies of workers were sent to check marshes and other watery places for larvae. In Queens and eventually in Manhattan itself, sprayers mounted on pickup trucks worked up and down the streets as police used loudspeakers to warn residents to get inside. Hospitals began to see a surge in emergency room visits for fever as nervous New Yorkers sought reassurance. Parents kept children indoors. Picnics and ballgames were canceled. So-called sentinel chickens were caged in mosquito-infested areas. It was believed that a change in their serology would signal the presence of the virus.

Many New Yorkers were gripped by mosquito fear. Their nervousness was no doubt heightened by frequent reports on the freakishly hot weather and oddities such as the news of two boys acquiring malaria at a summer camp on Long Island. (The parasite was imported by a visitor from abroad.) Eventually an edge of paranoia would be evident, as a local magazine published speculation about the outbreak being the product of a terrorist plot.

As more and more people became aware of the bird-mosquito-man cycle of SLE, dead crows became harbingers of fear. Officials told people to be on the lookout for the bodies, and even posted a telephone number to call so that health workers could retrieve whatever was discovered. Soon the phone lines were overwhelmed. Hundreds of birds were sent to the state lab. As some were confirmed to have the disease, the areas where they were found became subject to spraying. Of course, as the spraying widened, and intensified, many citizens started to question the safety of the insecticides that were being applied so liberally.

In retrospect, New York's response to the outbreak was remarkable for its rapid mobilization of significant mosquito-fighting forces. With few antimosquito resources of its own, the city managed to summon helicopters, spray trucks, and crews from outlying areas and put them to work quickly. Of course, killing mosquitoes with spray is far more complicated than one might think, and the particular mosquito that Mayor Giuliani hoped to attack was a more difficult target than most.

The problem was *Culex pipiens*'s preference for human habitats. Swamp mosquitoes can be dealt with efficiently through aerial spraying. But this mosquito likes to rest in enclosed spaces. In a borough like Queens, she will shelter under the eaves of houses and in garden sheds. In the concrete valleys of Manhattan, she'll seek out street drains, subway and utility tunnels, and even newsstands. She doesn't sit on a lamppost or the hood of a car, waiting for the pesticide truck to glide by.

Before they even sent out the spray trucks, city officials had to be sure that the wind was not so strong that it would blow the stuff away, or so weak that the cloud of insecticide would remain where it was deposited. They also had to wait until night, when temperature inversions—with cold air hanging above the warm streets—helped to keep the aerosol from dispersing too quickly.

When conditions were right, the trucks were turned loose, with the hope that the pesticide fog they created would kill enough mosquitoes to make a difference. Whether they succeeded will never be known. It is impossible to say that even a single case of encephalitis was prevented. That would require comparison of the pesticide with some sort of a placebo, and that would be unethical. And it is possible, given the life of *Culex pipiens*, that the mosquito was already fleeing the chill of autumn by the time the fog rolled in.

About all that can be said for certain is that the sight of the fogging trucks and helicopters played an important public relations role. It served the TV newscasters' need for images of the city's response to the crisis, and it alerted residents to the danger posed by the vector. In certain distant parts of the city, people rallied to demand that the spray be applied in their neighborhoods too.

On the negative side, the sight of pesticide wafting along city streets was sure to provoke a protest from people who feared the chemical more than they feared St. Louis encephalitis virus. This backlash is a predictable element in any outbreak in a modern city. Mothers worry that their children will be affected. Pet owners fret over their animals. Organic gardeners fear for their produce. In New York, criticism of the pesticide campaign began as soon as the trucks began to roll. But it would be cut short by a dramatic shift in the crisis.

In the last week of September, Bronx Zoo veterinarian Tracey McNamara realized that the actual virus found in the samples of bird tissues she sent to a national veterinary lab in Ames, Iowa, was simply too small to be St. Louis encephalitis. But while the

Ames lab could say that it wasn't SLE, it didn't have the resources to nail down the virus's exact identity.

McNamara turned to some friends at the U.S. Army Medical Research Institute of Infectious Diseases. They found that it was an African microbe called West Nile virus. The army's discovery of West Nile in animal tissue was accompanied, almost simultaneously, by its identification in human tissue at the CDC and at the Emerging Disease Laboratory of the University of California, which New York state officials had enlisted in their own investigation. The Connecticut Agricultural Experiment Station simultaneously registered the same discovery.

The news that West Nile virus had come to New York was played on the front pages of the newspapers. Public officials struggled to assure city residents that in a practical sense, nothing had changed. SLE and West Nile virus create roughly the same illness in roughly the same way. The city was still going to counterattack with adulticide, larvicide, and public health warnings. But in a historic sense, something dramatic was occurring. A brand-new, potentially fatal mosquito-borne disease had found its way to America from Africa and was threatening to take up permanent residence.

The CDC lost no time in their response to this outbreak. Responsibility for tackling the problem was given to their laboratory located in Fort Collins, Colorado. A group of scientists was immediately dispatched to New York City to define the threat posed by this novel introduction of an Old World virus into the New World. The director of the Fort Collins lab, Duane Gubler, became personally involved in the investigation. His own experience fighting dengue virus in Puerto Rico and in Asia prepared him for the problems that his people would soon face. Roger Nasci took up residence in New York where he led the entomological and epidemiological studies. Nick Komar, who had recently received his doctoral degree in my lab, led the bird-related studies. The New York State authorities appointed Laura Kramer to head their arboviral studies on West Nile virus, and she was joined by Greg Ebel, another recent product of my lab. Gubler served as the coordinating force that continues to guide the work of these research facilities as well as others in New Jersey, Connecticut, and Massachusetts.

The identification of the mystery virus explained the bird deaths and the human outbreak. It also offered a plausible explanation for the deaths of the horses on Long Island. Soon after the West Nile virus's presence was confirmed, the United States Department of Agriculture sent to Long Island an investigator from a high-security government lab on Plum Island. Doug Gregg traveled to a small farm where a horse named Terror had become a stumbling wreck and then suddenly died. A proud thoroughbred racehorse, she had just been taken off the track for good to be used as a breeder.

Before approaching the dead horse, Gregg donned a protective suit, complete with a hood and plastic face shield. Then he strode into a field and knelt beside the body. He took out a small hatchet and with deft little strokes cut a nearly perfect circle around the animal's ears. He lifted off a large, disk-shaped piece of the skull. After a little quick work with a scalpel he reached in and removed the horse's brain. In a matter of days it was determined that Terror had been killed by the newly arrived virus.

Horse deaths are a hallmark of West Nile virus. In Europe, Africa, and Asia the virus perpetuates itself in a cycle that requires birds to serve as its reservoir host. Except for certain geese, Old World birds do not seem to suffer much from the virus. Equine deaths have been noted in Egypt and among some wild horses in the south of France.

The virus was first identified in 1937 in Uganda near the western bank of the Nile Valley. Hence its name. In this region, WNV is just another common childhood disease. Virtually everyone is infected early in life. A moderate illness—fever, chills, muscle aches—may follow as the body fights off the infection and develops lifelong immunity. As with measles, the young are best equipped to resist, and the rare deaths that might occur would involve a child who is otherwise sick or weakened or an adult who somehow missed the immunity-conferring infection in childhood.

As we have seen, history is filled with the mortal results of encounters between endemic microbes of one region and nonimmune people from another. Whether the people go to the pathogen—as was the case with Mungo Park and malaria—or the virus goes to the people—as yellow fever did in eighteenth-century Philadelphia—the outcome is the same.

Compared with the major mosquito-vectored diseases, the history of West Nile virus's spread is relatively short. It was first seen outside its home range in the 1950s when a rash of encephalitis cases occurred among the elderly in Israel. The virus was later seen in many parts of Europe, in much of the Middle East, and in the western parts of Asia. More recently, significant outbreaks were noted in Algeria, Russia, and the Czech Republic. In 1996 and 1997, an estimated 50 people died in an outbreak in Romania. In Israel in 2000 more than 200 people were affected, with more than 40 deaths. In all cases, the spread of the virus seemed to follow the flyways of migrating birds.

Given the vast barrier of the Atlantic Ocean, how had the virus reached America? It's unlikely that we will ever know. But the possible scenarios are few. Perhaps the least probable involves an infected person landing at Kennedy Airport on a plane from some endemic area. Theoretically, mosquitoes could have acquired the virus from this person, but in truth, the virus level in the human bloodstream would be so low that such a transfer is almost impossible.

A second, more plausible, explanation for West Nile's arrival involves infected mosquitoes that stowed away on an aircraft and were released when the jet landed on American soil. Kennedy Airport sits in the middle of a wetlands region that teems with birds. The terminals themselves are home to thousands of house sparrows, starlings, and crows. The stowaway mosquitoes may have buzzed out of the aircraft with ravenous appetites, headed for the nearest bird, thereby beginning a cycle of transmission.

The third way that West Nile could have come to New York would have been via birds. Birds imported legally are carefully examined and quarantined, so they would not likely introduce a new pathogen. But a large illegal trade in birds thrives in the city and could have been the source of the virus. It is also possible that one or more wild birds managed the near miracle of crossing the Atlantic to import the virus in their blood. Birder lore is full of tales about species that are blown thousands of miles by the jet stream or a bizarre series of storms. Even a single bird transported this way to the New York area could spark an epidemic.

No matter how West Nile virus reached America, once it landed it could expand its range on the wings of almost any bird that flew. Crows seemed most vulnerable to the ravages of the disease. Thousands of them died, and during the outbreak they received a great deal of attention on the nightly TV news. But it is likely that the little house sparrow played the crucial role in propagating the virus. This bird is far more abundant in cities such as New York than are crows and tends to roost very near people's homes. The virus attains greater density in the blood of house sparrows than in crows and remains infectious to mosquitoes longer. The ultimate irony is that the omnipresent sparrow is itself a foreign species, which was imported by nineteenth-century lovers of Shakespeare, who desired to see every bird species from the Bard's era living in New York.

Fortunately for New York, the mosquitoes that apparently serve as the main vector for West Nile begin to look for a place to spend the winter as the daylight starts to recede in September. Although few in New York understood it at the time, the risk of new infections was approaching zero at the time the big announcement—it's West Nile virus—was made at the end of September 1999. Various agencies continued to look for the virus in dead birds, and it was found in a widening area that included parts of Long Island, other boroughs of the city, Westchester County to the north, New Jersey, and Connecticut. In all these places, people acquired the illness too. Serological studies suggested that a significant number of people—estimated to be nearly two thousand—were infected but felt mild symptoms or none at all.

Many aspects of life in the city were affected by the outbreak. People stopped dining outdoors and slathered themselves with repellents. When the virus was found in mosquitoes in Central Park, a free concert by the New York Philharmonic was canceled. Cops stood guard at every entrance to the park that same night while trucks equipped with foggers covered every acre with insecticide. The drama was so high that, at year's end,

Mayor Giuliani declared that protecting the citizenry from West Nile virus was the crowning achievement that year.

Tragic as they are, outbreaks of human illness provide an impetus for monitoring and for basic science that otherwise might go unfunded. A good example of this is the large research effort that has followed eastern equine encephalitis events in the northeastern United States. Triple E is far more deadly than either West Nile or St. Louis encephalitis virus. It kills about half of the people it infects. After six people were killed in the mid-1950s in Massachusetts, a system for tracking this virus was established, and it still functions today, protecting human health.

With the discovery of West Nile virus in New York, public health officials throughout the Northeast began systematic examinations of both birds and mosquitoes. They also began to look closely at the various kinds of mosquitoes that could maintain the virus in the host reservoir (birds) and transmit it to human beings. Though New York City had pinned the responsibility for both these activities squarely on *Culex pipiens*, the vector dynamic was far more complicated. The first complicating factor is that the typical *Culex pipiens* would rather not bite a human being, ever.

As it turns out, I had conducted intensive research on *Culex pipiens* at Harvard in the early 1960s. Each summer, I had hunted *Culex pipiens* in the utility tunnels beneath the Harvard Medical School complex, and the street drains in its vicinity. Every breeding site was mapped and the density of larvae recorded weekly. The entrance to the Medical School dormitory proved to be a most fruitful place to catch adults. I found them in the two phone booths that stood just inside the door and in the knee space under the desk where the guard sat all night long. He invariably left the outside door open to catch the breeze.

Many of the mosquitoes that I collected contained blood, and they were opened and the blood analyzed. House sparrows proved to be the main source of their blood meals. Indeed, human blood was present only during late fall, and not every year. The watchman rarely reported that he had been bitten, and the students appeared to be spared entirely. These mosquitoes were feeding virtually exclusively on the sparrows that nested and roosted in the ivy that covered the outside walls of the building. Although they maintained a raging epizootic of bird malaria (*Plasmodium relictum*) among these sparrows, we saw no need to try to eliminate these apparently harmless mosquitoes. Even the sparrows seemed to be thriving.

As part of the work, I reared new generations of mosquitoes from the blood-fed ones, and from those that contained no trace of blood in their abdomens. Those that could reproduce without blood were obviously *autogenous*. Typically they produced about eighty viable eggs. The blood-feeding or *anautogenous* mosquitoes were much more fertile, producing up to four hundred eggs.

In the laboratory, the two groups of mosquitoes were able to interbreed. Their offspring included a number—one-third—that were a kind of intermediate. The most remarkable thing about these intermediates was that they were far more likely to feed on human blood than their blood-sucking parents were. Apparently they inherit the desire for blood but not the instinct, if it can be called that, for avoiding human beings. Indeed, intermediates were present when those mosquitoes were biting the night watchman.

One expects that in the wild, the two kinds of *Culex pipiens* would interbreed only in unusual circumstances. In nature, the males of the blood-feeding variety form their mating swarms high in the air, often using fifty-foot-tall trees as swarm markers. Females fly up to encounter them, and it is at these heights that matches are made. In contrast, the autogenous *Culex pipiens* always find love at ground level, near the water's edge. In fact, upon emerging as adults, this variety of *Culex* doesn't move much at all before it begins the mating process.

Positioned as they are, fifty feet apart, how then do these mosquitoes ever produce hybrids? The answer may lie in stormy weather. If a rainstorm or heavy winds come along, they can drive the mosquitoes that are swarming aloft down to the ground. There, males and females of both types are likely to mingle and mate. When the autogenous males manage to inseminate blood-sucking females, hybrids result.

Storms can also contribute to population growth in *Culex pipiens*. The same squall that mingles the two kinds might fill stopped-up gutters, water barrels, and other sites for stagnant water. The water would be just right when the pregnant hybrids are ready to oviposit. When they ultimately emerge, the offspring fly off to seek the blood of both bird and man. Such hybrids may be the mosquitoes implicated in outbreaks of illnesses like St. Louis encephalitis. And it may be that weather conditions at breeding sites have something to do with the emergence of disease.

A thorough consideration of the bridge vector for West Nile leads to other suspect mosquitoes. Two—*Ochlerotatus solicitans* and *Aedes vexans*—live in New York marshes and are much more likely to bite human beings. *Culex salinarius*, which inhabits more permanent bodies of water, is also more partial to our blood. An attack on these mosquitoes might involve more targeted spraying in parks, wetlands, and open areas.

Ideally, an intervention would disrupt the cycle of transmission. An assault on the maintenance vector might protect birds from infection and thereby limit the virus's spread to humans. This might be done around the roost areas, where thousands of birds spend their nights and are often besieged by mosquitoes. If you choose instead to go after the bridge vector, you would try to kill those mosquitoes that transfer the virus from birds to people. In either instance, you would face a daunting task, as the virus takes flight on the wings of both birds and mosquitoes.

In the 1999 outbreak, we saw both the strengths and the weaknesses of the government's response to a new disease. The infections and deaths among the elderly were noticed early and investigated. It seems clear, however, that the laboratories, which screen for just six pathogens, could have subjected the samples that were sent to more thorough testing.

The big question that was posed after the West Nile outbreak of 1999 was: Would the virus survive the winter in the New World? The answer was soon provided by its discovery in the bodies of several female *Culex pipiens* hibernating in a tunnel in an abandoned fort near the recent epicenter of the outbreak.

In spring and summer, mosquitoes and birds were examined from Texas to Maine. By autumn the virus would be detected in a dozen states, from New Hampshire and Vermont southward to North Carolina. New York State, with perhaps the most thorough surveillance system in place, found the virus in 1,271 dead birds, 360 pools of mosquitoes, eight live birds, two sentinel chickens, two bats, one squirrel, and one chipmunk.

Severe human illness was centered in New York and New Jersey, and the number of confirmed cases—twenty in all—was

much smaller than it had been in 1999. Interestingly, the epicenter of the outbreak shifted away from Queens across New
York City to Staten Island. Just two people died. Both of them
were in their eighties.

West Nile virus also erupted in the Middle East. Numerous
human infections were recognized in Israel during the summer
of 2000. West Nile virus is genetically variable, and the genetic
variant that was responsible for the North American outbreaks
in 1999 and 2000 was identical to the virus that was isolated in
Israel in 1998. One has to wonder if a peculiarly virulent version
of West Nile virus has emerged worldwide.

In the United States, horses suffered much more from the
virus than did people in the year 2000. Some sixty-five horses
came down with severe neurological symptoms caused by the
virus. It is interesting to note that horses continued to fall ill with
the virus in October, after human infections abated.

There is no doubt that birds are carrying West Nile virus up
and down the eastern seaboard. We should also expect to see
the virus creep westward. In the year 2001, it may well arrive
in Pittsburgh, Cleveland, or even Chicago. The effect on human
beings is more difficult to predict. Why did fewer people get
West Nile encephalitis in 2000 than in 1999? The answer may
be that people avoided mosquitoes. It may be that spraying destroyed important vectors. Or it may be that the weather was a
significant factor. The summer of 2000 was unusually dry, and
human-biting mosquitoes were therefore scarce. The number of
human cases of mosquito-borne disease was small: one case of
EEE compared with an average of five. Three cases of SLE instead of 124. Zero western equine encephalitis when a normal
year produces 18.

Of course, myriad variables make a definitive analysis of West

Nile's behavior in 2000 impossible. In other parts of the world, illness caused by West Nile virus appears almost randomly. Outbreaks occur, then many years may pass before the disease comes back. This dynamic may be explained in part by the way a viral infection immunizes a bird. Bird populations can eventually become immune to the virus. For this reason, viruses may circulate in birds across a broad region, moving to find new generations of birds that have never been infected. A pathogen that starts out, say, in Belgium, may blaze its way southward over several years before it circles back to find newly vulnerable populations of birds in Brussels.

West Nile virus is now rooted in the North American landscape and will give us good reason to swat at mosquitoes for the foreseeable future. It may spread across the northern and middle tiers of states as far as the West Coast. It is not clear whether the virus will invade the deep South. Mosquitoes there may not be able to maintain it in birds or serve as the bridge to humans. But these questions are just now being asked, and the answers are in the future. After all, no one predicted that a mosquito-borne virus from Africa would colonize America.

Other viruses lying in wait "out there" might, some day, follow West Nile's route. Japanese B encephalitis is perhaps the most likely prospect for the invader role. This disease is far more severe than West Nile virus, and people of all ages become ill. The list of additional candidates is long. It includes Ross River and Murray Valley viruses from Australia, and Rift Valley virus from Africa.

# 9

# LIVING WITH MOSQUITOES

Today much of the tropical world suffers from chronic problems due to malaria. In many places, conditions conspire to create particularly devastating outbreaks. For example, in January 2001 heavy rains and political unrest that disrupted antimosquito programs were blamed for a fivefold increase in malaria in northern Burundi. More than 275,000 cases and 115 deaths were reported. Drug resistance was a major problem.

Other mosquito-borne illnesses are flaring up all over the globe. Ross River virus, which causes a rash and a flulike illness, is proliferating in Australia. Honduras recently experienced an outbreak of dengue that killed at least ten people. Approximately 190 deaths were recorded in a burst of yellow fever in northwestern Guinea. North America braces for another summer with West Nile. And the Middle East is grappling with Rift Valley fever, which killed more than two hundred people in the year 2000.

These outbreaks and epidemics highlight our continuing struggle with mosquito-borne disease. The great conundrum, however, in our relationship with the mosquito is that much of the trouble is caused in part by our own activities. These fall into three main categories:

First, there is the problem of international travel and commerce. West Nile's arrival in New York is only the latest example of a dynamic that will be repeated. Mosquitoes and pathogens are now global citizens, and we should not be surprised to see them show up almost anywhere, anytime.

The second cause is our mishandling of the medicines and insecticides that now are available. Indiscriminate and haphazard use of these chemicals has spurred the evolution of resistant strains of disease agents and vector insects. This woeful process continues worldwide. Our best antimalaria drug, chloroquine, is losing efficacy. And the most powerful arrow in our antimalaria quiver, DDT, is being outlawed.

A worldwide ban on DDT would be a mistake. When properly used, DDT can be uniquely helpful, especially in less developed countries where public health funding is exceedingly restricted. But the ban imposed by the United States and fierce advocacy by organizations devoted to environmental improvement have severely restricted its use. DDT is a prisoner of politics and may never escape. If it is banned worldwide, human lives would be placed at risk while this tool, which is safe when used properly, could save them.

The third self-imposed problem is our failure to attain goals that we set for ourselves in the past. Our failed attempt to rid the world of malaria, and to keep South America free of *Aedes aegypti*, has damaged public confidence. In the wake of these defeats, public health workers experienced a loss of direction, policymakers became skeptical, and funding agencies began to experience "donor fatigue."

Despite the weight of previous failures, and the variety of threats that we endure, we can learn to live with mosquitoes. The practical means to do this will be different in rich countries and poor, in the Northern Hemisphere and in the Southern. But if we have the will, we can find a way to share the planet safely.

In the industrialized world, living with mosquitoes now means accepting that an outbreak of disease might occur at any time. And just as nations guard against biological terrorism, they now must conduct surveillance against naturally occurring agents that are delivered, not by human enemies, but by blood-feeding insects.

A recent example comes from North America, where environmental monitoring has become the centerpiece of the public health defense against West Nile. There public health officials have focused on dead and dying birds. Birds are the reservoir hosts for the virus, and certain birds will suffer first whenever West Nile moves into a new neighborhood. Because West Nile could pop up anywhere in North America, it would be wise for many cities and states to examine dead birds for the virus.

Of course, any local wildlife pathologist would dread being overwhelmed with the carcasses of thousands of dead birds. This is exactly what happened in 2000 to New York's state lab, when citizens sent in every dead bird they saw.

A well-designed surveillance system would develop as risk of an outbreak intensified. It should also avoid overwhelming the labs. Initially, reports of virus-infected dead birds (especially crows) would provide the crucial warning that West Nile virus is present in the region. After a few virus-infected crows have died in a site, however, the discovery of another such animal would contribute little additional information. Instead, surveillance should focus on virus-infected mosquitoes and particularly on the kinds of mosquitoes that bite human hosts. Various kinds of traps are available for this purpose. The presence of numerous human-biting, virus-infected mosquitoes demonstrates an urgent need for acute public health warnings and for interventions directed against adult mosquitoes. However belated, reports of two or more human infections in a site would be even more dire.

Along with monitoring birds, virtually all major cities and states should have well-tuned systems in place to signal an impending outbreak of a novel human illness such as that caused by West Nile virus. The "passive case detection" systems that are already in place throughout the developed world depend on physicians to report to central monitoring facilities certain categories of disease in their patients. This might be augmented by a program designed to monitor diagnostic laboratories, not just for positive findings, but also for sudden increases in requests for certain kinds of tests. Even when such tests return negative results, the very fact that physicians have increasingly ordered certain kinds of tests may be meaningful.

As soon as an outbreak is noted, health authorities must act quickly to identify the source of disease and to attempt to prevent new clusters of illness. Probably the first step in all outbreaks is public notification. This must be done with care. In the early 1960s, I spoke to a reporter about discovering *Culiseta melanura*—which can carry eastern equine encephalitis—at Harvard Medical School. Though there was no evidence that the mosquitoes were infected with the virus, the reporter's newspaper immediately published an article announcing "Brain Fever at Harvard." In the uproar that followed, I feared losing my job.

Today a more careful mosquito expert and a well-informed reporter would, one hopes, do a better job of placing a real outbreak of something such as West Nile in its proper perspective. The public should be alerted but not frightened. Sensible precautions—covering up, avoiding prime biting times outdoors—can be issued without suggesting a cataclysm.

Once human disease has struck, agencies like the Epidemiological Intelligence Service of the Centers for Disease Control and Prevention can provide the scientific manpower to determine where the virus lurks. (Most developed nations have services similar to the CDC's.) Citizens can be told how to avoid mosquitoes by covering up and limiting outdoor activities. But ultimately local authorities will have to decide whether to attack the mosquito vector with pesticides, and whether to go after its adult aerial stage or its aquatic larval stages.

The decision is difficult, and public pressure on those who must choose will be powerful. Again, New York's experience with West Nile is very instructive. There, initial anxiety about the disease spiked so high that even a vocal group normally opposed to putting any chemicals in the environment actually pushed for pumping pesticides into all the city's subways, sewers, and utility tunnels. But early calls for widespread application of pesticides were followed by outcries over the dangers of the spraying. A fight over pesticide safety inevitably occurs once fear of the virus moderates.

Though whipsawed by citizens on either side of the issue, politicians will have to consider the scientific advice available to them, and this evidence may not be at all conclusive. In the case of *Culex pipiens* in New York City in the 1999 outbreak, the mosquito would have already begun its seasonal retreat, seeking its winter hiding places, by the time the outbreak was recognized. Such knowledge might have influenced the health authorities to hold back on spraying. Of course, the situation is much more complicated. Some *Culex pipiens* do not hibernate at all, but breed continuously throughout the winter in sheltered sources of water such as basement sumps. And the oldest mosquitoes that emerge before the onset of hibernation are most likely to be infectious.

Another major concern in the spray debate revolves around the means for delivering the chemical. There is little doubt that aerial spraying, from helicopters and low-flying airplanes, produces the most uniform attainable insecticidal coverage. Airplanes are especially effective over wetlands and parks where roads are widely spaced and few. The insecticide blankets the region evenly and generally reduces mosquito abundance substantially. The effect on mosquito density may not immediately seem obvious if new adults continue to develop. But the troublesome Methuselahs will be gone.

Truck-mounted sprayers are not so effective. Buildings block the dispersal of the spray, and coverage tends to be incomplete. Even so, in densely populated areas trucks provide the least intrusive method for delivering mosquito-killing fogs. People naturally fear having their homes, pets, and children doused from above.

Because the mosquito identified as the main West Nile vector is peridomestic, the New York health authorities couldn't enjoy the luxury of sending helicopters to less-populated areas and spraying swamps. *Culex pipiens* doesn't live in swamps. It lives in cities and towns. To mount a credible attack, New York had to fight in the city streets, and this meant less-effective, truck-mounted sprayers. (Cities would be sprayed from the air only when the impending outbreak would be truly calamitous.)

In New York the decision to spray seemed to be correct. Truck spraying is often the only thing that can be done. It will kill at least some mosquitoes, and it does assure local residents that the outbreak is being taken seriously. When used properly, the chemicals present only minuscule risk to the environment. Though one can never prove that a particular case of encephalitis was prevented by a spraying program, in many cases the possible benefits outweigh the possible risks.

As an alternative to a spray attack on flying mosquitoes, the health authorities dealing with any threatened outbreak may attempt to reduce the abundance of a vector mosquito by attacking its larvae. The available "larvicides" include such microbial insecticides as *Bacillus thuringiensis israelensis* (known as *Bti*) and *Bacillus sphaericus*. These microbial preparations appear to be uniquely safe because their toxins act solely on mosquitoes and their close relatives. They destroy the wall of the larva's gut. A synthetic juvenile hormone preparation, such as methoprene, is nearly as safe. This preparation interferes with metamorphosis. A larvicidal attack on a mosquito-borne disease is more difficult because each breeding site must be identified. In the case of the common house mosquito vector of West Nile virus, this requires treatment of each street drain, bird bath, clogged downspout, sump, abandoned swimming pool, rain barrel, and blocked ditch in a neighborhood.

Besides guarding against new pathogens, specialists in mosquito-borne disease must be on the lookout for immigrant species of mosquitoes. These insects are remarkable travelers. Many survive transoceanic flights at high altitude hidden in the baggage compartments of jet aircraft. Container shipping offers further opportunities for mosquito travel, and mosquitoes even hitch rides on trains and in motor vehicles.

Though we may rightly fear that some exotic mosquito will arrive in our neighborhood bearing a pathogen in its body— and this may be what happened when West Nile virus was carried to the New World—this isn't the only way a new mosquito can cause trouble. Sometimes the new insect is capable of spreading pathogens that are already present in the environment but previously had no way to get to humans.

Consider the consequences of the introduction in the Northeastern United States of another small-container breeding mosquito like *Aedes albopictus*. If it focused its bites solely on large mammals such as deer and human beings, this interloper might transmit Jamestown Canyon virus or Cache Valley fever, caused by viruses that are already in many deer, to people. The list of pathogens already present in our environment is long. The threat posed by some new vector and vector-borne infection is so complex that the situation defies prediction.

In addition to any actions that public health agencies may undertake, individual residents of an affected site can protect themselves, at least partly, against mosquitoes and the infections they transmit. Some knowledge of the kinds of mosquitoes that are present where you live might help you to avoid their bites. It is useful to know that:

The main West Nile vector, *Culex pipiens*, feeds at night and breeds only in foul, stagnant water. A sump in the basement of your house may be a source where these mosquitoes breed. So too are clogged gutters that retain water. Anything that might hold water—disused tires, bird baths, old cans, rain barrels, abandoned wading pools and swimming pools—require attention to limit breeding. Finally, air-conditioning and screens in good repair probably do more than anything to keep you safe from *Culex pipiens*.

*Anopheles* mosquitoes, which transmit malaria, quest for blood mainly after dusk and just before dawn. A few strays may feed at other times of the night, but none will be around during the bright, sunny parts of the day. These mosquitoes breed in relatively permanent, clean water. They can thrive in any temperate-to-tropical climate, from Europe to South Africa.

Eastern equine encephalitis and other encephalitis viruses are transmitted in the United States by daytime feeders, such as *Aedes vexans, Ochlerotatus sollicitans*, and *Coquillettidia perturbans*. These three have different habits. A late-summer biter, *vexans* breeds in temporary rain pools. (There's little we can do about these sites.) *Vexans* are sneaky and attack after you have been still for a few minutes. More aggressive, *sollicitans* breeds in salt marshes and will literally chase you to get your blood. They are abundant after heavy onshore winds and a spring tide. They are horrendous biters. *Perturbans* breed in freshwater, cattail swamps. They are almost as annoying as *sollicitans*. It is possible that West Nile virus may also be transmitted by these mosquitoes. All feed during the day and at dawn and dusk and mostly out-of-doors. Few bite during the brightest, hottest times of the day—an hour or so around noon.

If you live in or visit a tropical place where dengue is endemic, you should be alert to the habits of *Aedes aegypti*, which is strictly a dawn and dusk feeder. They bite either indoors or out. This mosquito breeds in small, artificial containers of water, such as flower vases and disused automobile tires.

To avoid most of these mosquitoes you can stay away from

their breeding areas, and time your outdoor activities for moments when they are least active. Various precautions can be taken when you do go outdoors. Long sleeves and pants limit the skin available for biting, and some repellents do work. Those containing the chemical DEET are effective, and while nerve damage can occur after massive overuse, they are safe when used properly. (Avoid formulations that contain more than 34 percent DEET, because this level of pesticide can increase the chances of an adverse reaction.)

It's interesting to note that almost any oily substance, such as mineral oil, thickly applied, will discourage mosquitoes from biting. (This may be why some skin care products seem to repel mosquitoes.) DEET-containing formulations last for many hours after the carrier they are mixed into has dried.

Multitudes of plant-derived compounds have been investigated as repellents. Some have seemed effective. Citronella works, but if it's in lotion form it must be reapplied frequently. "Mosquito coils," which are pyrethrum-containing candles, will help reduce the number of mosquitoes at your barbecue. The insecticide they waft into the air is derived from chrysanthemum flowers. A related synthetic compound, permethrin, is used by the United States military to impregnate combat uniforms. Other people may take similar protective measures where mosquito annoyance is intense. This chemical is the active ingredient in many commercial household sprays. But it should never be inhaled or applied directly to skin.

What doesn't work?

- Electronic bug zappers that attract and electrocute insects. These kill mostly moths and very few mosquitoes.

- So-called ultrasonic repellents that supposedly drive mosquitoes away with low- or high-frequency sound.
- Culture of natural predators, such as birds and bats. They just don't eat enough mosquitoes to make a big difference.

As countries in the industrialized world seek to exclude mosquito-borne disease from their borders, those in South America, Africa, Asia, and the countries of Oceania seek to rid themselves of the various endemic diseases. Malaria, in particular, is becoming increasingly widespread and in many places increasingly resistant to the medicines that we may use against it. At the same time, the various anopheline mosquitoes that transmit this infection are becoming immune to our diminishing stock of available chemical insecticides. Over vast regions, human health suffers and national economies fail due to this resurgent affliction.

Determined to find some way to deal with malaria- or dengue-bearing mosquitoes, the leading research-funding agencies—including the Tropical Disease Research Section of the World Health Organization, the Wellcome Trust of Great Britain, the MacArthur Foundation, and the U.S. National Institutes of Health—have focused many of their resources on the idea that the natural vectors of malaria might be replaced by mosquitoes that are genetically modified so that they cannot transmit this infection.

Basic science has gained much from this attempt to produce a genetically modified mosquito. The genomes for certain mosquitoes are almost fully mapped, and sets of genes that make a mosquito incompetent have been identified. Tony James at the University of California at Irvine and Alex Raikhel at Michigan

State University have successfully introduced genetic markers into mosquitoes in the laboratory. "Genetic drive mechanisms" have been discovered that might help a particular gene become dominant when an engineered mosquito is released in nature and mates with the native population.

The goal of manipulating genes in order to render a mosquito vector impotent is exceedingly attractive. Instead of destroying our enemy, we'll gently convert her. She'll continue to buzz and bite but will never kill again.

It is likely that these research efforts will succeed in combining a gene that prevents a pathogen from developing in a mosquito, with a drive mechanism that will carry that gene through a caged population of mosquitoes. And the media will likely herald the arrival of "The Answer" to malaria.

But it seems equally certain that in the real world, in the countries, cities, and villages in which malaria and *Anopheles* mosquitoes do their damage, this approach will fail. The realities of both nature and humankind conspire, in many ways, to defeat the kinds of interventions planned by molecular biologists.

The challenges begin with the genes themselves, which might be delivered via a transposon—an element that serves as a kind of carrier for a gene. There is no guarantee that in the wild, the genes and the transposon will stay connected. Such transposons tend to jump and genes to mutate. But even if they remain together, the so-called good mosquitoes would have to surmount a great many obstacles in order to help us.

First, the laboratory-bred mosquitoes used to introduce the "engineered genes" into nature may fail to compete for mates because they have become adapted to life in a cage.

Then again, we may have to nurture our artificial infestations of modified mosquitoes and maintain them at densities greater than occur naturally. Insecticides may have to be outlawed, and bed nets eliminated, at least until certain goals are attained. Imagine how people living near these mosquito colonies would respond to a huge population of biting mosquitoes, against which they are helpless to defend themselves.

Finally, the genetically engineered insect may confront different kinds of isolated vector mosquitoes and different mosquito-borne pathogens in a single village. Some of the vector mosquitoes will be unable to mate with the released mosquitoes, and other diseases will continue unabated.

All of these problems will confront this high-tech genetic scheme, even before one considers the complicated human and political dynamics that greet anyone who ventures into the field to fight mosquitoes. A good example of such difficulties is captured in my encounter with a man who called himself Big Knife.

It happened on Grand Bahama Island. I was traipsing through villages looking for discarded tires, bottles, chamber pots, and other containers that might harbor larval mosquitoes. I stayed well back from the street, moving along the fence line that marked the limits of each property. Behind one especially well-littered house, I found myself facing a very large man who challenged my presence near his property. A discussion of mosquitoes and the diseases that they transmit was fruitless. When he asked, "Guess why they call me Big Knife?" I fled.

Other Big Knives living throughout the world will resent anyone mucking around in their backyards. And they certainly will resist the idea that "good mosquitoes" are going to be released near their homes. This was tried before, in India, when the World Health Organization launched a mosquito abatement effort involving the release of male house mosquitoes that had been rendered sterile by radiation. They could inseminate females, but their sperm would effectively sterilize any female that accepted their advances.

Almost immediately the Indian program was met by violence when the residents of the site misunderstood this attempt to sterilize mosquitoes and thought that it was aimed at themselves. The American Central Intelligence Agency was said to be sterilizing people. The project was soon canceled, never to be resumed.

It may simply be impractical to release a genetically modified mosquito. Local residents will object to any increase in mosquito bites. Political activists will seize the opportunity for confrontation. Safety issues will arise if other mosquito-borne pathogens are present in the site. Ethical issues involving "informed consent" and "right of withdrawal" may be insuperable. Given the rocky history of genetically modified foods, fiercely opposed in many quarters and banned in many countries, what chance is there that people will accept a genetically modified mosquito?

For nearly twenty years, those who set the agenda for science funding have taken an optimistic view of the myriad obstacles to the genetic approach to disease-bearing mosquitoes. They feel that the power of modern science is limitless. The practice of molecular biology is wonderfully exciting, involving ingenious experiments based on technology that can, at times, seem almost magical. But laboratory work is reductionist. It bears down on ever smaller aspects of a problem. In this case it has ignored two important obstacles to success: Big Knife and nature.

Although efforts based on molecular biology continue to consume an overly generous share of the funds available for scientific research on vector-borne diseases, the World Health Organization just launched a major attempt to reduce the burden of malaria in the world. The World Bank, various United Nations agencies, and many governments have joined in this effort to reduce the hardship that malaria imposes on tropical countries. The campaign is called Roll Back Malaria, with the goal of reducing mortality by half by the year 2010. Called RBM for short, the program is described as a stool that stands on three legs: distribution of insecticide-treated bed nets, clinical care based on combinations of drugs, and intelligence-based interventions.

Insecticide-treated bed nets will serve to reduce, but not eliminate, exposure to malaria-infected anopheline mosquitoes. Imperfect as they may be, bed nets can ease the severity of the illness. This is because the nets reduce the number of infectious bites that a person receives each night. Superinfection thereby becomes reduced, and this helps a person's immune system battle the malaria parasite.

The second RBM strategy, medical "case management," is complicated by the rising problem of drug-resistant malaria. The old standby drug, chloroquine, now fails about half the time in much of the tropical world, and the second-line drug, Fansidar, seems about to succumb to resistance too. The objective for today's health care practitioner then is to design a first-line treatment, based on combinations of drugs, mainly artesunate followed by mefloquine. The theory holds that this combination will forestall any adaptation by the malaria parasite.

Combined therapy will involve an enormous increase in cost, and it is hoped that the developed nations of the world will help. The scheme is now subject to intense debate. Some argue that delivery of two drugs at the same time will expend both as rapidly as if only one were used because many people will not follow instructions, particularly if they are illiterate and poor. Some may make errors in using the drug. Others may share medicine with friends and family. They could, on their own initiative, revert to one-drug therapy and undermine the whole treatment concept.

The third RBM strategy—intelligence-based interventions— seeks to deliver insecticides or drugs with precision, but only when needed. This sophisticated strategy requires the use of computerized maps, known as geographical information systems, or GIS, that integrate masses of information derived from different sources. Meteorological information, records of malaria diagnoses, drug use, and reported deaths are some of the types of information that can be used.

The GIS system would compare the present situation with the history in each community in the region. A mobile "fire brigade" would be ready to travel to an affected site and to undertake the antimalaria activities that seem appropriate. This strategy would seek to recognize unusual circumstances and would be based on a concept of malaria containment rather than suppression. A group of South African scientists has established such a system (known as MARA) for all of sub-Saharan Africa. My own attempt to establish such a system in Eritrea in 1996 was interrupted by war and was only recently resumed by others.

Although every element of the WHO's program carries a flaw or two, one shouldn't assume that RBM won't work. In fact, the agency is likely to meet its goal and Roll Back Malaria deaths by one-half in a decade. Millions of people will be saved, and 2010 will likely see a celebration of this very real victory.

We must hope, however, that the victory is not transient. The drug distribution measure will be effective as long as anti-malaria drugs can be purchased and resistance is held in check. As long as bed nets are impregnated with insecticide in a timely fashion and mosquitoes remain susceptible to these chemicals, lives will continue to be saved. If sudden outbreaks can be detected in a timely fashion and effectively truncated, then even more people will benefit.

But the most intractable problems will be faced after RBM attains its goal and a new basis is required for sustaining these improvements in health. The health authorities of impoverished nations may not be able to assume this responsibility. Any loss of first-line drugs and insecticides would require substitution with more expensive chemicals, and the local authorities may not have the needed funds. And any ultimate reversal of a successful antimalaria program might be disastrous because millions of residents of malarious sites would then have been rendered nonimmune. With malaria in place in the environment, such immunological virgins will present highly vulnerable targets to malaria parasites.

For these reasons, new ideas are needed to take us beyond the RBM strategy to a permanent fix. Some of them might be found if we approach the problem from a fresh perspective. Instead of fashioning a new chemical attack on the mosquito or the malaria parasite, we might follow the protocols used in places where malaria disappeared without a chemical war on the insect or the pathogen. It happened in the United States, with the Tennessee Valley Authority, in Italy with Mussolini's agricultural revolution, and in nineteenth-century Africa, where the British and the Germans did create malaria-free communities.

In each of these situations, general improvement in the local economy produced better housing, roads, and utility services such as water supplies, sewers, and electricity. At the same time, elementary techniques were used to separate people from mosquitoes. Mainly, this meant improving the quality of housing while eliminating nearby breeding sites.

To do this today in the tropical world we need a program of research that aims to create fundamental changes in the relationship between human and mosquito populations. For example, at Harvard efforts now focus on local malaria problems in Ethiopia and Namibia. My colleagues are trying to understand how human activities may promote malaria transmission. One of these studies flows out of basic research on the breeding habits of the vector mosquitoes.

The larvae of the *Anopheles gambiae* complex of mosquitoes, the most important of the African malaria vectors, generally are found in the muddy water that collects in the "borrow pits" that are left after people dig up soil to make mud bricks for their homes. Because so much silt remains suspended in this water, it was difficult to understand what the larvae could be eating to grow so well. The answer lay in the pollen from the corn gardens that seem to surround every African home during the rainy season.

To test this idea, some breeding pits were covered with raised tents of clear plastic that would exclude most pollen grains. Pollen was added to similar pits located some distance from a corn patch. Time and again, pollen yielded a bumper crop of large adult mosquitoes while its absence meant starvation and death for the larvae.

Because mud brick homes must be rebuilt almost continually, Namibian villages have lots of these little water-collecting borrow pits. And because people don't like to carry these heavy bricks very far, such pits generally are found right next to homes. To protect their crop against thieves, corn is planted as close to the home as possible.

The malaria equation seems obvious. Mud bricks plus corn equals disease. Corn and mud bricks are essential to the lives of these people; but that doesn't mean we can't intervene. One solution might be to distribute seeds of genetically engineered corn that would produce pollen that is toxic to larval mosquitoes.

The genetically modified corn solution, however, might not be acceptable in today's climate of suspicion of transgenic foods. A complementary approach might be to reconsider the way that houses are built in Africa and to promote the use of a better housing material that might reduce the number of borrow pits in a village. Perhaps one could promote the use of bricks made of rammed earth, which last much longer than traditional mud bricks but require a centralized manufacturing site. A primitive cement mixer is required because such bricks contain a trace of portland cement, and a simple press is needed to finish the optimal product. The eaves of such houses can be closed and the structure made insect-proof with a few simple window screens and doors.

The construction of this kind of house does more than protect its inhabitants. It begins to address the structural problems of poverty. A permanent house is valuable property. The owner of a home is likely to invest in the structure as though it were money in the bank. Upkeep maintains the health benefits of a house. Its value may also give the owner financial leverage to invest in better farming techniques, or even education.

But the benefits of home ownership do not flow immediately. In much of Africa, home ownership is rare, and left-leaning politicians resist private ownership of land. With effort, however, communities can be persuaded to try this approach. Eventually the well-being of an entire village, or town, could improve. Certainly anything that promotes health and development is needed in these places. After all, they are faced not only with malaria, but with HIV/AIDS, which is killing many people in their prime earning years. Families left behind when wage earners die contribute to the drag on local economies.

Naturally, outside funds will be needed to confront malaria in poor countries, but eventually these nations will have to solve this problem themselves. If the international community were to help, it might need to target its aid in a rather unconventional way. This means helping, not just the poorest of the poor, but also those who are nearer to what we'd call a middle class, the richest of the poor.

The idea would be to spur development by breaking the isolation that countries suffer when they are plagued by malaria. This can be done if a few stable, malaria-free sites were to be created that would attract investment and perhaps tourism from the outside world. Mosquitoes would be "built-out" of contact with people. Such clean cities would become both magnets for workers and engines for further growth. A self-sustaining cycle of health and wealth, independent of foreign aid, would be extended to more and more people, as well as to more and more communities.

The key to our relationship with the mosquito is getting to know it better. After more than a century of serious inquiry, great mysteries remain. We still don't know what it is about blood, specifically, that mosquitoes crave for reproduction. We don't know how mosquitoes distinguish among hosts. How does a female *Culex pipiens* know that it is approaching a bird, and how does an *Aedes aegypti* distinguish a person from a horse? We don't even understand something as basic as how a mosquito develops the sucking force necessary to pull blood through a feeding tube with a diameter that is so infinitesimal that the resulting friction should make it impossible. The salivary tube is even finer. How does the salivary pump work? Why do certain kinds of mosquitoes support the development of one kind of pathogen and not another?

The beauty of the mosquito lies in these mysteries and many others. Posed against an enormously dangerous environment, this seemingly simple organism thrives. Everything about its design is economical and precise. And though it is incapable of thought, it manages to meet great challenges, adapting to our use of pesticides, the loss of habitat, even climate change. Charles Darwin would have been amazed at the speed with which those mosquitoes that exploit the human environment adapt and diversify today.

In their enormous variety, and their adaptive genius, the ubiquitous mosquito provides us with an up close view of nature's evolutionary logic. For those mosquitoes that live with us and depend on us for food, their dependency seems to be their strength. Vector mosquitoes, like weeds, thrive best where their environment is disturbed, the condition that characterizes the human footprint on the land. If we were to disappear, many mosquito species would likely become extinct. Many others, however, would carry on. Mosquitoes were here on earth before us and will likely survive us.

With their glassy wings, delicate legs, and seemingly fragile bodies, mosquitoes are nevertheless a powerful, even fatal presence, in our lives. What creature could be more intriguing?

# WHAT DID I JUST SWAT?

A small insect flying around a person would almost always be a female mosquito, particularly if it has at least one long forward-pointed projection on its head (the proboscis) and a pair of feathery antennae. If you swat it lightly (or spray it with insecticide), its remains generally can be identified with the aid of a hand-lens and a strong light. The form of its cranial projections and the pattern of dark and light markings on its body, wings, and legs provide useful clues to its identity. These characteristics can readily be seen with a magnifying glass.

The "big three" of the mosquitoes that transmit human disease worldwide include various *Anopheles* species, *Aedes aegypti*, and *Culex pipiens*. Although dozens of different mosquitoes may be present in a single site, these common mosquitoes can often be distinguished from each other by considering the following characteristics:

> *Anopheles* mosquitoes, the vectors of human malaria, are brownish and have three long, similar-looking projections on their heads, a proboscis (bundle of feeding stylets) and

two equally long palps (sensory appendages). When the mosquito is at rest, the palps are held against the proboscis; they are raised when the insect is feeding; and they generally become splayed outward in a dead anopheline mosquito. The palps should not be confused with the fine, feathery antennae that normally are held outward. White patches are present on the wing-veins of many of the more dangerous anopheline mosquitoes. The ends of their abdomens are truncate.

*Aedes aegypti* mosquitoes, the vectors of dengue and yellow fever, have very short palps that are too small to see without a microscope. Their bodies appear jet black and are ornamented with silvery white lines and patches. Their wings are clear and translucent. Their abdomens end in a point.

*Culex pipens* mosquitoes, the vectors of West Nile virus and filariasis, have very short palps. Their bodies are browish overall, but have transverse white lines on the upper surface of their abdomens (hind bodies). Their wings are clear and transparent. The ends of their abdomens are truncate.

# CHARACTERISTICS OF ANOPHELINES AND CULICINES

Kent S. Littig and Chester J. Stojanovich

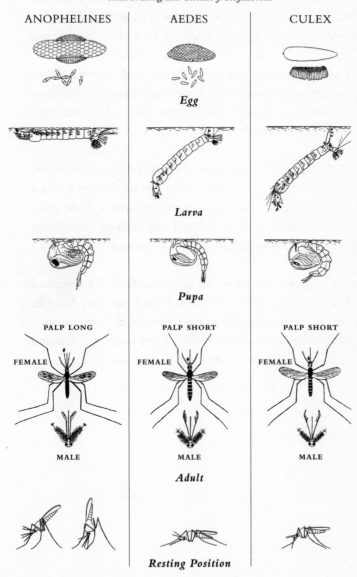

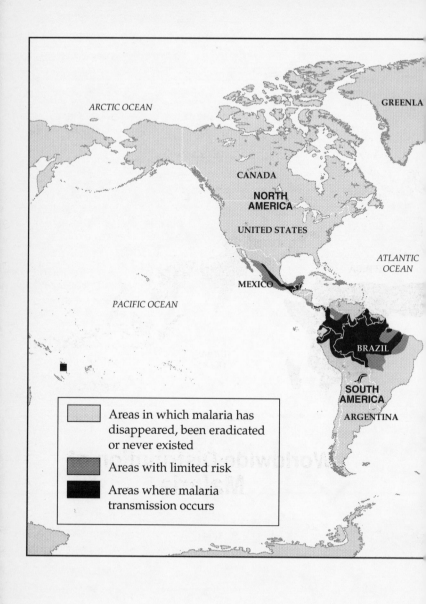

Areas in which malaria has disappeared, been eradicated or never existed

Areas with limited risk

Areas where malaria transmission occurs

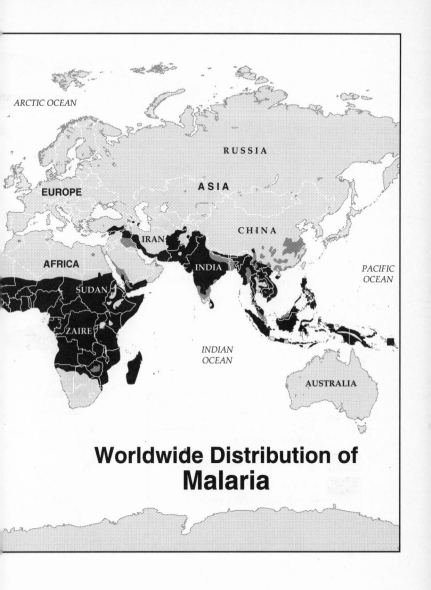

# Worldwide Distribution of
## Malaria

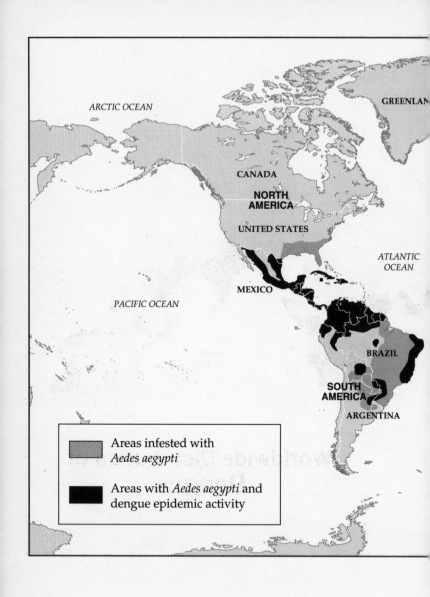

Areas infested with
*Aedes aegypti*

Areas with *Aedes aegypti* and
dengue epidemic activity

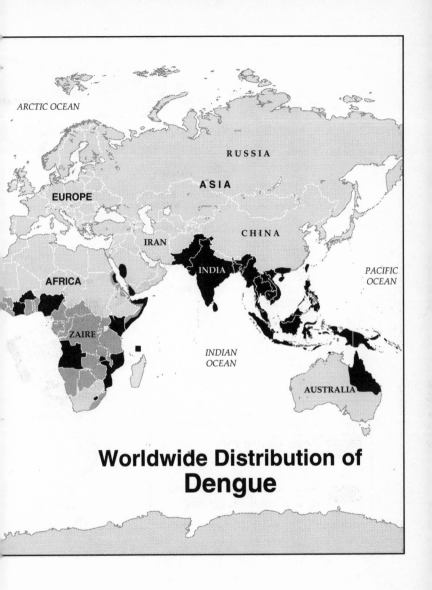

**Worldwide Distribution of Dengue**

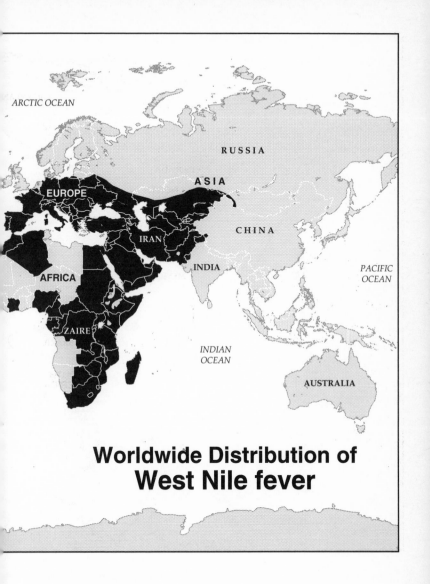

**Worldwide Distribution of West Nile fever**

# INDEX